ANIMAL KINGDOM

A NATURAL HISTORY IN 100 OBJECTS

JACK ASHBY

The History Press

First published 2017

The History Press
The Mill, Brimscombe Port
Stroud, Gloucestershire, GL5 2QG
www.thehistorypress.co.uk

British Library Cataloguing in Publication Data.
A catalogue record for this book is available from the British Library.

ISBN 978 0 7509 8152 1

Typesetting and origination by The History Press
Printed in Turkey by Imak

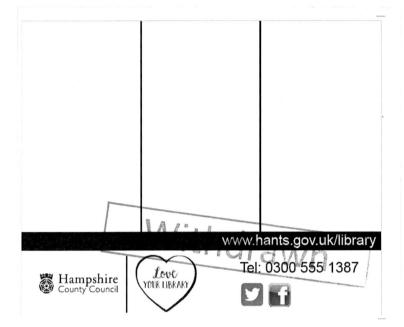

Contents

Introduction

My home contains many skulls, fossils, rocks, shells, bones, and teeth. Most are reminders of trips I've taken and things I've seen – natural souvenirs from my life as a zoologist. The same items also exist in museum collections, but those are not souvenirs. When objects are put in museums they become specimens, imbued with some intrinsic and intangible value they did not have before they arrived there. They become exemplars of a group – representatives of their species or kind, setting a standard for a particular animal from a particular place at a particular time. By contrast, I think of my objects at home more as knick-knacks, there for my own enjoyment. They can tell stories, but they are just my stories.

Objects in museums tell different stories. And what we can learn about the animal kingdom from museum specimens is also very different from what we can learn watching animals in the wild. In my ecological fieldwork in Australia I am lucky enough to have engaged with thousands of animals at extremely close proximity. These kind of encounters – and the scientific research they are part of – help us understand how ecosystems work and how animals behave in the wild. The stories we gain from these experiences enable us to see nature in action, as a vibrant, impossibly complex series of living interactions.

Encounters with specimens from dead animals in museums give us a different view. These meetings happen on our own terms, and allow us to ask and answer different questions. I have seen many more living Tasmanian devils than I have dead museum specimens, for instance, but I feel like I have learned more about them, or at least different things about them, by close study of their remains in museums. Natural history objects allow us to investigate the evolutionary history of animals, and seek to find not just what they do, but how they do it.

Research in the wild and in the museum are both critical pillars of natural history; neither tells the whole story on its own. We can only go so far in studying a skeleton or a carcass to imagine how the animal is adapted to moving through its environment, for example. Similarly brief encounters with animals on their own terms do not tell us everything we seek to know about where a species came from and how it came to be.

Museum specimens also allow us to study animals that have been dead for centuries. New technologies are developed to answer questions that weren't even imagined when the animals were first collected. Examining fossils allows us to understand species that have been dead for millions of years. We can use them to build a picture of worlds long disappeared, and how the earth came to be the way it is now.

Objects can tell the general stories about their kind, but also the specific stories that they alone were characters in. They are at once specimens *and* individuals – museum objects were collected, traded and used by people: their stories are historical as well as scientific. Investigating museum collections tells us something about the time in which they were collected. They provide windows into the past as well as data for the present. These historic repositories represent centuries of ecosystems and cultures. Even though they are just dead animals, they reflect local, national and global politics, and the societies they are linked with.

The UK alone holds well over 100 million natural history objects in its public museums. The global collection runs into the billions. The scale of these numbers is impossible to conceive of, and this is just one of the reasons why only a minuscule proportion of museum collections are on public display – usually less than 1 per cent. The functions of objects on display are very different to the collections held in storerooms, and this is one thing that I'll be exploring in the final section of the book.

Animals are everywhere in the human world: they are used to advertise pretty much everything, from banks to toilet paper, and cars to cereals. They are mascots for sports teams and logos for airlines. The way they are represented usually links to some reflection of their natural history. As a result every single person walks down the street with an extraordinary knowledge of nature, even if they don't think so themselves. This means unlocking some level of understanding of the natural world is easy. Museums can help us access some of this ingrained expertise, by sparking connections between what people already know and the amazing, rare objects in museum collections.

What I have attempted to do with this book is to select 100 museum specimens to tell 100 stories from the animal kingdom: 100 objects, 100 stories. It is not a history in the sense that it doesn't have a beginning, a middle and an end, and on the whole it doesn't run chronologically. The animal kingdom doesn't have one history. From a single beginning, countless millions of histories have – and continue to – run their course. They intertwine constantly, some of them end abruptly and some fizzle out.

The book is arranged into four sections, each telling a different kind of story. The first explores the diversity of life over the past 600 million years or so, which is mainly the story of the invertebrates (animals without backbones; vertebrates are animals with backbones). Although information about natural history is everywhere in the human world, the coverage is not even. Large and charismatic species get the lion's share of the limelight. Even though they outnumber vertebrates by more than twenty to one, invertebrates are given a raw deal in natural history museums, television programmes and books. I'll admit that I am also somewhat guilty of this (as mammals are my passion), but the first section of this book attempts to show most of the ways there are to be an invertebrate as I go through many of the major divisions of the animal kingdom.

The Turning Points section tells the story of evolution from the perspective of us mammals. As I explain, this is only one of the countless routes through the branches of the tree of life. It begins with the sudden appearance of most of the key groups of animals a little over half a billion years ago, and explores the appearance of different kinds of fish. Our evolutionary history involves fishes without jaws and without hard skeletons, a huge diversity of bony fishes, the colonisation of the land, the end to a reliance on reproducing in water and responses to massively changing ecosystems. We can trace our relationships with fishes, amphibians and reptiles, and understand how the world's mammalian fauna came to be the way it is today.

The third section is by far the biggest. It explores the animal kingdom by investigating some of the most exciting mechanisms in evolution. By understanding some of the key stories of how nature operates, we can gain an amazing insight into the workings of the world. The objects in this Natural Histories section reveal some of the astonishing systems underlying life. Nature is truly incredible, and with the objects in this section I have tried to highlight some of the genuinely astounding ways it works.

Finally, in the Displaying Nature section I hold a lens up to museums, one of the key windows we have onto the natural world. I ask questions about how the decisions that have been made in the creation and curation of museums throughout their history influence the way we see animals today. Museums are not sterile or natural places, but human inventions. As such the biases of human nature and politics can affect our interpretation of nature.

All of the objects in this book come from the collection which I manage – the Grant Museum of Zoology at University College London. However, in truth they could have come from any museum with a natural history collection. Aside from the individual historical stories that objects can tell, and perhaps a few genuinely rare 'hero objects', to some level of approximation all natural history museums have examples of more or less the same stuff, it's just that the collections vary in size. I could have chosen any number of objects to illustrate these 100 stories, as every species and every individual is a result of its deep ancestry, subject to the same evolutionary pressures as everything else.

To set the scene I have chosen one object to kick things off, before we get into the four separate sections. I wanted to illustrate how the book works, with an object and a story. It explains one of the key concepts that underlies how evolution works, and uses my favourite animal to do it.

1 Platypus

TAXIDERMY SPECIMEN

Evolution Can Only Work With What It Has Got

Science is supposed to be unbiased, but scientists are not. We have our favourites and our passions, and mine is the platypus. I could have used this bizarre Australian to illustrate any number of nature's mechanisms or incredible stories in the history of zoology, but I wanted to use it to demonstrate on a very simple idea about how evolution works, as a means to begin the book: Evolution can only work with what it has got.

One thing that evolution had at the end of the Triassic period, a little over 200 million years ago, was the synapsids – a group of reptile-like animals which would give rise to mammals. Triassic synapsids had a lot of characteristics that we might call reptilian. This is because they shared a recent evolutionary history with the groups that would become crocodiles, dinosaurs, lizards and turtles. They did things that reptiles do, like laying eggs and walking with bent elbows and knees, with their limbs held out at right angles to their bodies (or at least partially).

This is the ancestral frame that evolution had to work with. All new developments would have to start from here. This means that the first mammals would have shared a lot of characteristics with their reptilian forebears.

Platypuses are often described as 'primitive' because throughout their long evolutionary history they have retained these ancient ancestral features. Platypuses, and their closest living relatives the echidnas, still walk with bent elbows and knees, and still lay eggs, just like their Triassic relatives.

Platypuses are mammals, and laying eggs might not seem like a very mammalian thing to do. Many people are under the impression that all mammals give birth to live young. There are a few unique defining characteristics of all living mammals, but live-birthing is not one of them. They include the presence of hair, three inner-ear bones, a specialised ankle joint and a single bone forming either side of the lower jaw (reptiles have seven), and the ability to produce milk (mammary is the root of the word mammal). Platypuses have all of these features. (Platypuses do not truly suckle their young, though, as they don't have nipples. However, they do have mammary glands: the milk kind of sweats out of them, and the babies lap it up.)

Together platypuses and echidnas form the order Monotremata, so named for their cloaca: a single (*monos*) hole (*trema*) for all their urogenitary functions. Like reptiles, they do all their defecating, urinating, mating and birthing through this one single opening. It is these reptilian features that have earned platypuses their 'primitive' reputation. However, on to this ancient template platypuses have added some of the most advanced characteristics of any mammal. This is the reason why I get so excited about platypuses.

First, male platypuses are among the only venomous mammals. They have a horny spur on their heels attached to a venom gland, which is used during sexual competition. (The venom is said to be excruciating and long-lasting for humans.) The males of many species have developed myriad ways to secure sexual success, from giant antlers to absurd decoration. Platypuses are the only mammal to use venom to fight off rivals.

Second, they are one of only a tiny number of mammals known to be capable of electroreception (the others are their close relatives the echidnas, and one species of dolphin): they can detect electricity through their bills. Platypuses are restricted to the lakes and rivers of Tasmania and eastern Australia, where they are reasonably common. They use their power of electroreception to hunt worms and tiny crustaceans. Muscular movement in animals is controlled by electrical impulses. Little is known about platypus feeding, but it is assumed that platypuses can locate their prey buried under silt by sensing these electrical impulses with their bills. They gather food up in little pouches at the base of the bill and then mash it up (adult platypuses have no teeth) with their rubbery bills when on the surface.

In the history of the platypus, evolution started with a 'primitive', reptile-like template. Without changing much of this ancient frame, incredibly advanced features were added. This demonstrates that a modern *species* can never itself be called 'primitive' – only the *characteristics* it has inherited from its ancestors. Neither can a species be considered 'advanced', as although it may have evolved some brand new tricks, it still retains those basic elements that have stood the test of time and are found to be still useful now.

When Europeans first discovered platypuses at the end of the eighteenth century they caused quite a stir. The first specimens to reach England – preserved in barrels of brandy or as dried skins – were thought to be hoaxes, made from at least three animals sewn together. There are (probably apocryphal) stories that the earliest specimen to reach what is now the Natural History Museum in London has visible plier marks on its bill, where the curator tried to wrench it off to demonstrate it had been sewn together as a fake. Indeed, in the Grant Museum of Zoology, where I work, there is a bill that has been detached from its body in a crude manner that implies it cannot have been carefully dissected away, which may be the result of similar treatment.

With its bizarre mix of characteristics, both novel and ancient, the platypus did not fit into the existing animal groups that scientists had conceived, and there was great debate over where to place it. Eventually, entirely new groups were created to accommodate it.

Taxonomy aside, it took nearly a century to settle the argument over whether it truly did lay eggs. Perhaps the religiously minded scientific elite took exception to the idea that something which was clearly a mammal could do something so reptilian – something that might drag our noble class down into the mud and slime.

Robert Grant – the founder of the Grant Museum and an early proponent of evolution (he had a huge influence on sparking Darwin's thinking) – was a strong supporter of egg-laying in platypuses. Perhaps he used this taxidermy specimen in his search for evidence.

Part 1

Understanding Diversity

The human desire to classify is perhaps at its strongest when it comes to natural history. From our childhood years we are taught to put the animals we encounter in museums, living rooms and the natural environment into discrete categories. At school and on television we are taught the differences between groups like amphibians and fish.

Thomas Henry Huxley, one of the greatest biologists of the nineteenth century and a man who can take a lot of credit for making Darwin's work accepted by the Victorian scientific community, said this:

> To a person uninstructed in natural history, his country or sea-side stroll is a walk through a gallery filled with wonderful works of art, nine-tenths of which have their faces turned to the wall. Teach him something of natural history, and you place in his hands a catalogue of those which are worth turning around.*

He was right. Knowing what you are looking at – in both nature and the art gallery in Huxley's analogy – is half the joy. But we can't become 'instructed' in everything. The world is an endless purveyor of wonders too numerable to memorise. To make sense of the 1.5 million species so far described (and there may be 100 million undescribed species), natural historians have to come up with a system for arranging them and information about them.

This proved tricky until 1735, when Swedish botanist Carl Linnaeus proposed a system for putting species into hierarchical groups, and it stuck. In today's terms, a rat can be a rat, a rodent, a mammal, a vertebrate and an animal all at once. Such taxonomic thinking is really important for how we understand the world and our place in it, as each of these terms comes

---

* Huxley, T.H., *Lay Sermons, Addresses and Reviews* (D. Appleton: New York, 1876)

with implicit information about how they relate to other groups. It neatly puts the world into boxes. Although it certainly wasn't Linnaeus' intention (he believed that studying nature would reveal the divine order of God's creation), hierarchical taxonomies tell us a lot about an animal's evolutionary history, as by their nature they show what came from what. This information is real and truthful, but, as is often said, a bee doesn't care that it is a bee. Taxonomy – the science of putting things into groups – is a rigid human construct that is forced on top of the cacophonous uncertainty of the real wild world.

One of the central tenets of modern taxonomy is that every group has to include, by definition, all of the groups that evolve from it. So rats did not stop being mammals when the rodent group branched off the evolutionary tree. Every branch on the tree of life is considered to be a member of all its parent branches.

This means, for example, there can be no definition of fishes that does not include everything that evolved from fishes. Following this logic, you could argue that as amphibians evolved from fishes, amphibians are fishes. Mammals evolved from animals that evolved from fishes, so mammals are fishes. We are fish. While every biologist knows this conundrum, and that there is no biological definition for what most people consider 'fishes', they decide not to worry about it because it's helpful to think about living swimming 'fishes' as a group. Taxonomy is useful and makes a lot of sense, until it doesn't.

Similarly we have all been taught that the animal world can be divided into vertebrates and invertebrates. This is a very handy division, but it suffers from the same problem as 'fishes': as vertebrates evolved from invertebrates, every single species of animal there has ever been is an invertebrate. Given that this means that animal and invertebrate are taxonomically synonymous, we just have to agree to ignore that and carry on as normal.

Left: Comb jelly illustrations from *Kunstformen der Natur* (*Art Forms of Nature*), a 1904 book by German biologist Ernst Haeckel (see object 18).

The boxes that taxonomy forces around the natural world come in a set sequence of ever-more specific groupings. The largest in common use is kingdom, and this book focusses on just one of them – the animals. The simplest hierarchy of groups goes like this:

Kingdom
Phylum
Class
Order
Family
Genus
Species

And for us, that hierarchy would look like this:

Kingdom: Animal
Phylum: Chordate
Class: Mammal
Order: Primates
Family: Great apes
Genus: *Homo*
Species: *Homo sapiens*

Below the kingdom level, the next sub-grouping is phylum (plural: phyla). There are commonly held to be thirty-five animal phyla. This first section of the book makes an attempt to give coverage to some of the diversity of life that doesn't enjoy a lot of limelight. I could have chosen to dedicate a chapter to each of the phyla, but I haven't.

That's because evolution has produced many different ways of being a worm: about half of the phyla are distinctly wormish. I apologise to worm-fanciers out there, but given that many of these phyla contain rather few species, and that I have a lot more to say about the less vermicular animals of the world, I opted not to give them the proportional coverage you may think they deserve.

Instead, here follow eighteen different objects representing eighteen of the major groupings of the animal kingdom. It hopefully demonstrates most of the key ways there are of being an animal.

A few of which are worms.

2 Elephant Ear Sponge

DRY SPECIMEN

Sponges: The Poriferans

For most of the history of life all organisms were single cells. A giant leap was made when dividing cells didn't separate from one another and became integrated to form a simple multicellular organism. Only then could huge steps be made to diversify the shapes, sizes and behaviours we see in the biological world today.

Sponges are among the oldest known animal fossils, dating back to the late Precambrian period, 580 million years ago. They branched off extremely early in the history of animals, separating them from the

groups that have far more complex structures. They have one of the most simple body plans of all animals. They only have a small number of different types of cell, and these are not organised into properly differentiated tissues. They do not have any organs.

Because they are so unlike all other animals, until the 1820s sponges were believed to be plants. Robert Edmond Grant, the founder of the Grant Museum of Zoology, was a radical thinker who subscribed to 'transmutationism': the idea that species change over time – what we now call evolution. Grant spent several years studying and publishing on sponges and other simple invertebrates. He believed that they could provide insights to the roots of the tree of life.

Much of what we now know about sponges is down to Robert Grant. He noted that they passed out a fluid waste, which was evidence that they were digesting something – a key defining feature of animals. He also found that although sponges were not known to move, their reproductive cells could. Together this helped prove that sponges were in fact simple animals, not plants. He also came up with the name that unites all sponges into a single phylum: Porifera – 'the animals that bear pores'.

Sponges live in all aquatic habitats, from the poles to the equator. They are extremely successful filter feeders, built as a network of channels lined with cells called 'collar cells'. These have a single hair-like flagellum

(like the tail on a human sperm) that they beat to form a current, sucking water through pores, along the channels and then out through one or a few large exit holes. In a single day a sponge can pass an amount of water equivalent to 20,000 times its own volume.

Sponges come in three classes, distinguished by the crystal structures called spicules that hold them together and

Venus' flower basket – one of the glass sponges that are made almost entirely of silica.

essentially form their skeletons. Demosponges are the largest group and includes the elephant ear sponge, pictured on p. 17, and bath sponges. Demosponges have either fibres made of a collagen protein called spongin, spicules made of a silica mineral, or both. Calcareous sponges have crystals made of calcium carbonate, and the glass sponges have spicules made of silica crystals (but different to the ones demosponges have). This last group can form beautifully regular lattices of silica – particularly in Venus' flower basket sponges (opposite) – which makes it difficult to picture where exactly the 'wet' part of the animals' cells go. They are animals that are naturally formed of 90 per cent glass.

In the nineteenth century it was discovered that sponges' collar cells are strikingly similar to a group of single-celled organisms called choano-flagellates. This has led many zoologists to believe that choanoflagellates are animals' closest relatives. In this case, sponges would be the most primitive group of animals, having evolved from a choanoflagellate-like ancestor.

Sponges do not have many of the features seen in other animals: they are not symmetrical in any way, and as well as lacking organised tissues they don't have muscles, sense organs, nerves or a mouth. They don't have respiratory, circulatory or kidney-like systems, but all parts of their bodies are very close to water so they can do fine without them. Gases and waste products can simply diffuse in and out using their water pumping system. It's hard to rule out the possibility that sponges' simplicity has evolved from more complex animals. Rather than never having evolved them, instead they may have lost some of these key features because living passively on the seabed does not require a very complex way of life. Interestingly, sponges have been found to have the genes that in other animals control some components of nerve impulses, but sponges do not have any nerves.

Because of the simplicity of sponge structures, they can reproduce simply by breaking apart. A fragment knocked off by waves or fishes has everything it needs to grow into a new sponge. They can reproduce asexu-ally (without involving reproductive cells) by budding off a mini-replica, or by producing a gemmule. This is a little package of cells packed into a solid spore-like structure. Gemmules are particularly common among freshwa-ter sponges which are more susceptible to drying out or freezing. Sponges can also reproduce sexually, and most are hermaphrodites, i.e. each indi-vidual can produce both male and female reproductive cells. Some species release eggs and then sperm into the water separately, so they can fertilise or be fertilised out in the open water. Others just release sperm, which are then trapped in other sponges by their filter feeding mechanisms.

3 Sea Anemones

TRAMOND WAX MODELS

Jelly With A Sting: The Cnidarians

Many of the world's jelly-textured animals are grouped together in the phylum Cnidaria. This name takes its root from the Greek word for nettle – a reference to the fact that all cnidarians use stinging cells for hunting and defence. Anemones, corals, jellyfish and hydras are all cnidarians. They are a diverse, global group of over 12,000 species, all of which are aquatic (apart from some extremely unusual parasitic forms), but only twenty of them live in freshwater.

Wax models of snakelocks anemone, mottled anemone and *Scolanthus callimorphus* made by the nineteenth-century Parisian anatomical model-makers Maison Tramond.

Hard corals, like this brain coral, produce stony skeletons out of calcium carbonate that they extract from the water.

Cnidarians are the only animals visible from space. You can't make out individual corals from a space station, but together millions of them have built Australia's Great Barrier Reef – the world's biggest living structure. Corals are responsible for creating entire ecosystems upon which untold numbers of other organisms rely. Coral reefs are some of the most diverse spots on earth: most cnidarians may be small, but they have a truly global impact.

Reefs are built of calcium carbonate structures deposited by corals which can take on a huge diversity of forms, from round corals, flat corals, branching corals and pipe corals. Each 'lump' of coral is built by hundreds or thousands of tiny individual coral animals called polyps that form colonies of clones, stuck together and working together as if they were a single animal (see chapter 68 for more on colonial living). Many species of coral have photosynthetic algae living inside them, producing nutrients for the coral in return for a safe place to live.

Despite the range of shapes that coral structures take on, up close the individual polyps display the features that make them recognisable as cnidarians. The corals, sea anemones, hydras and jellyfish are united by having a mouth (which also functions as their anus) surrounded by tentacles – with stinging cells called cnidocytes – and radial symmetry: when looked on from above the same pattern is seen if they are rotated.

The typical cnidarian life cycle alternates between two stages: a sedentary polyp (picture an anemone), which is basically a hollow tube fixed to a solid surface at one end, with a mouth at the top; and a free-living medusa phase (picture a jellyfish), which has formed by flattening the polyp's tube into a disk, but with the mouth at the bottom. Some cnidarians have mixed life cycles with phases as each, some are only ever polyps, and some are only ever medusae.

Soft-bodied cnidarians, unsurprisingly, do not have a particularly recognisable fossil record. Definite, hard cnidarians are known from 400–500 million years ago, but many people have speculated that some of the most enigmatic and oldest animal fossils known are cnidarians, from the Precambrian Ediacaran period, well over 600 million years ago.

Along with sponges, comb jellies and the microscopic plate-like placozoans, cnidarians sit outside of the main grouping of animals, the Bilataria, which contains everything else. Bilatarians, named for their bilateral symmetry (they are roughly symmetrical around a midline), are animals with a front and back and top and bottom. Their sensory organs are concentrated around the mouth in a head (a feature not seen in cnidarians, which have a mouth but no head), and the digestive tract empties out of an anus – an evolutionary step up from using one's mouth to defecate.

In any case cnidarians and the comb jellies represent a step up in animal complexity from the sponges. Cnidarian cells are organised into true tissues, but they do not have any organs, which are structures built from a combination of tissues working together. They have muscles, nerves and senses, as well as a mouth and a digestive cavity. As such they are more closely related to humans (and the rest of the Bilataria) than they are to sponges, and together cnidarians, comb jellies and bilatarians form a group called the Eumetazoa.

Cnidarians do not have circulatory or excretory systems, and no central nervous system. Their nerves are arranged into simple networks that control the contractions of their muscles and process the information from their sensory organs (which can include the detection of chemical signals, pressure, gravitational pull and light – see box jellyfish eyes, chapter 39) This may sound like a rather simple arrangement, but it has worked extremely well for the Cnidaria; they have survived for hundreds of millions of years and are critical elements of many aquatic ecosystems.

The key characteristic features of the group – the nematocyst stinging structure – are an amazing feat of engineering. When triggered, a coiled harpooning filament from within the cell fires into the victim at incredible speed – around 2m per second. This is astonishing considering it is generated by just part of a single cell. They then release a potent neurotoxin that acts on the nervous system to paralyse the victim. When this tactic is used in hunting rather than defence, the muscles in the tentacles then contract to pull the paralysed prey towards the mouth.

4 Comb Jelly

MICROSCOPE SLIDE

Comb Jellies: The Ctenophores

Pickling and other means of museum preservation do not do justice to this group of gelatinous, fragile and beautiful marine invertebrates. In life they are cloudy sacs of jelly covered in bands of fine hairs called cilia. The way that their cilia refract light as they beat in synchronous waves produces iridescent rainbow flashes, and furthermore they can even produce their own light from bioluminescent cells within their bodies.

Comb jellies form the phylum Ctenophora. They are small – most are just a few centimetres long, though they can range in size from just a few millimetres to just over a metre – and there are only 100–200 described species across all marine environments, but due to their role in the food chain as both significant predators and also as prey they have an important ecological impact on their habitats.

Aside from the larval stages of a couple of species, all comb jellies are predators. They swim very slowly, by beating their eight 'combs' of cilia down the length of their bodies. They are in fact the

Microscope slide of the comb jelly *Callianira sp.*

The comb jelly *Beroe sp.* feeds mainly on other species of comb jelly.

largest animals to swim using cilia. Moving like this, without using any kind of paddle or flipper, means that they can get very close to their prey without alerting them by disturbing the water in a way that a conventional swimmer would.

Many species hunt using a pair of tentacles covered not with stinging cells, like cnidarians, but with little spots of glue. When an animal gets stuck on these they unwind a long spiral thread that ensnares the victim in a tangled gluey trap, which the comb jelly then reels into towards their mouths.

Others simply engulf their prey by shutting their mouths around them, and in a few species special cilia in the mouth are used as 'pseudo-teeth' for biting gelatinous prey such as other comb jellies. Their mouths open into a large digestive cavity where they break down their prey. Unlike cnidarians, which just have a mouth, comb jellies have small anal pores at the other end of their bodies. Some waste is ejected through these tiny openings, but most of it just comes back out of the mouth.

This is one of the many lines of evidence that have been used to try and establish where they fit on the tree of life. It is possible that the anal pores of comb jellies are evolutionarily linked to the true through-gut that bilatarian animals have (where all food comes in the mouth and all solid waste out of the anus).

It's unclear whether the anal pores do suggest that comb jellies represent another step in increased complexity in the evolution of animals. Because of the simplicity of their bodies, the paucity of useful fossils of the right age and the uncertainties of extrapolating genetic comparisons over half a billion years or more, it's very difficult to make firm conclusions about the early stages of animal evolution. Often the different tools used by evolutionary biologists suggest different answers to one another.

Comb jellies almost certainly fall outside the huge animal group that is the Bilataria, but their exact relationship to that group and to the cnidarians and the sponges is highly contentious. Because of their similarities to cnidarians they have often been placed in a group together, but these features would appear to have evolved independently and these two groups of jelly-like animals are not closely related.

Some people believe that they may be the very earliest group to split off on the first animal branch of the tree of life (even before sponges and the microscopic plate-like Placozoa). When the comb jelly genome was mapped, scientists at the University of Florida found that they lack many of the genes that are critical for the development of nerves in all other animals. This is rather surprising: comb jellies do have nerves, but now we don't know how they build them. They were then found to lack all but one of the neurotransmitters that other animals use. The authors of the study claim that this suggests that nervous systems in comb jellies evolved separately from all other animals, meaning that nerves evolved twice.

Many evolutionary biologists have found this hard to believe, as the steps that were required to develop a network of communicating nerves were extremely complex and unlikely. For them to have appeared twice in the history of life is even more unlikely.

It's incredible that any fossils of these bags of fluid have ever been found. However, there are definite comb jelly fossils from 380 million years ago, during the Devonian period, but the group definitely predates that. There are some possible contenders in the famous Burgess Shale of Canada (a fossil site that contains species that represent some of the earliest members of many major animal groups (see chapter 20)) from around 500 million years ago, but they differ slightly in structure from modern species.

All in all, a final answer on how these simple creatures fit into the story of animals is still beyond our grasp.

5 Parasitic Flatworm

Flatworms: The Platyhelminthes

The phylum that includes flukes, tapeworms and free-living flatworms – Platyhelminthes – contains worms with a wide diversity of lifestyles and adaptations. Flukes and tapeworms are parasitic, with some species living inside their hosts, some on the surface. Many species need to infest more than one species to complete their lifecycles and go through a very variable number of different developmental stages along the way. Free-living flatworms live in water or moist environments.

The features that unite them all are not easily spotted. When discussing this phylum nearly every phrase comes with the line 'in some species'. They are restricted to aquatic environments: water, the wet films coating soil particles and leaf litter, and the wet internal environments of other animals, as parasites (around 80 per cent of the group are parasitic).

Platyhelminthes are collectively called flatworms, and their flatness is an adaptation to aid the transfer of gases across the body wall to and from the outside (which is why they are restricted to damp habitats – water is required to dissolve oxygen and carbon dioxide in order to cross cell membranes). This is necessary, since they possess neither circulatory nor respiratory systems.

The majority do not have a gut either. This is a common adaptation associated with parasitism – internal parasites live in a world opposite to our own – they are bathed in their food and absorb nutrients from the outside of their body; their 'skin' is adapted accordingly. By contrast we absorb food from within our bodies, via our guts.

The guts in the species that do have them are one-way bags. The mouth is often located in the centre of their underside, and food comes in and waste goes out of that same mouth. Sometimes their stomachs come out of their mouths too: many free-living species can turn their stomachs inside out to swallow up prey outside of their bodies. Within their bodies, the gut can have many extending fingers branching off of the centre so that nutrients can reach every part of the body. Rather than being allied with those animal groups that fall outside of the Bilataria which also lack an anus, platyhelminthes are believed to have decreased in complexity in their evolutionary history, and lost the through gut secondarily.

Free-living species swim by beating their cilia and by muscular undulation of the body, rather like a flying carpet. They are predatory, feeding on insects, molluscs and other worms. They have a range of adaptations for capturing prey, including wrestling them by wrapping their muscular bodies around them, coating them in a gluey or toxic mucus and stabbing them with their penis-sword. They may begin to digest their prey outside their bodies by secreting enzymes, before sucking up the resultant soup.

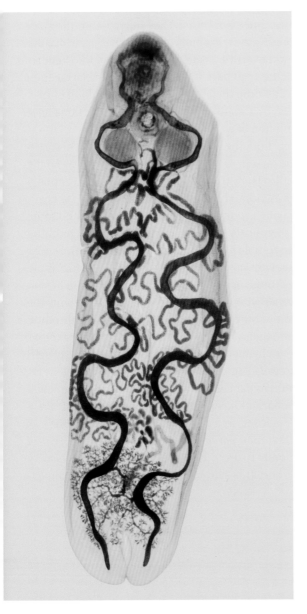

Tools for adherence are commonplace among platyhelminthes. Some have structures called duoglands which excrete a glue to stick them to surfaces to stop them from being washed away by fast currents (they also exude a solvent to dissolve the glue to allow them to move). Some have vacuum suckers to attach themselves to prey in the case of predators, or hosts in the case of parasites. Tapeworm heads have complex hooks and suckers to lock them into the wall of their host's gut. Aside from that part, nearly the whole tapeworm body is dedicated to reproduction.

Microscope slide of the fish parasite *Isoparorchis sp.* – a diginean flatworm.

The chances of making it through all the stages of their multi-host life cycles are slim, so parents have to stack the odds by creating thousands of fertilised eggs to ensure that at least some of their offspring survive to reproduce themselves.

These life cycles can be astonishingly complex. The liver fluke *Fasciola hepatica*, for example, has to go through stages living in a snail, a human (or sheep or cow) and freshwater in order to reproduce. Eggs are passed out of a mammalian host in their faeces, and if they end up in water they hatch into a larval stage. They then actively seek out specific species of lymnaeid snail to act as intermediate hosts. They burrow into the snail and metamorphose again into a cyst that buds into multiple larvae of a different kind, which feed on the snail's tissues. Once matured into another life stage, they bore out of the snail back into the water. These larvae then swim to some vegetation and drop their tails, forming a new kind of cyst. There they wait to be eaten by a mammal (many human infections have been linked to the consumption of watercress).

Once inside the mammal host, the cysts burst to release juvenile worms which burrow through the gut, on to the liver, and eventually move to the bile ducts where they settle. Here they sexually mature and eventually reproduce. Fertilised eggs are ultimately passed out in the faeces to begin the cycle again. Other parasites can require three species of host, and some modify the behaviour of their hosts to increase their chances of reaching the next 'level' (as with the zombie death-grip fungus, chapter 59).

Platyhelminthes have a wide variety of sexual tactics too. Some of the free-living flatworms are famous for 'penis fencing' – attempting to stab their hermaphroditic partner with their penis-swords: whoever wins the swordfight gets to inject the other with sperm. Many hermaphroditic parasites release sperm into their host's system to be absorbed by another member of the same species living there (though they are capable of self-fertilising if no mates are available).

Schistosoma, a major human parasite responsible for the disease schistosomiasis (also called *bilharzia*) has male and female forms, but once paired the female lives enclosed in a groove along the male's body. A fish flatworm parasite *Diplozoon paradoxum* has taken relationships a step further and permanently fuses to its partner sharing a body cavity with no obvious dividing line. It is said to be the world's most monogamous relationship.

6 Bootlace Worm

Ribbon Worms: The Nemerteans

What is it about worms that so many people find distasteful? A sea slug (see chapter 58) that is brightly adorned with vivid stripes of green, yellow, blue or purple would be considered beautiful, or even cute. So why when a similarly patterned animal is elongated into a worm-shape does it become monstrous? Many ribbon worms – members of the phylum Nemertea – are wormy browns and beiges, but there are a lot of extraordinarily patterned ones too. Like the sea slugs, these bright colours signal to would-be predators that these species carry nasty toxins and it would be unwise to attempt to eat them. Such pigmented warnings are called aposematic colouration and are discussed in depth with other objects later on (see chapter 42).

One nemertean from the North Sea is purported to be the world's longest known animal: the 55m-long bootlace worm. Specimens of this length have been reported but not been formally registered in museum collections, but 30m examples are regularly found (for reference the largest recorded length for a blue whale is 33.5m). Worms of this size are pretty astonishing considering they are only 0.5 to 1cm thick.

When disturbed, bootlace worms can quickly retract themselves over metres of surface to shelter in a crevice. Being so extendable obviously makes measuring them difficult. Despite being named exactly 99 years after the birth of the father of modern taxonomy Carl Linnaeus, in 1806, its scientific name *Lineus longissimus* appears not to be a misspelled homage to him. It's hard to know whether Linnaeus would be disappointed by that.

Most ribbon worms are highly capable predators or scavengers, found mostly in the sea but also in rivers, lakes and damp environments on land. Bivalve molluscs, annelid worms and crustaceans fall victim to their unusual feeding techniques. The group is characterised by a muscular trunk-like proboscis that can reach almost the length of their entire bodies. It is shot out in order to catch food. The proboscis can simply grab and wrestle their prey into the mouth, or coat their victims in mucus string and

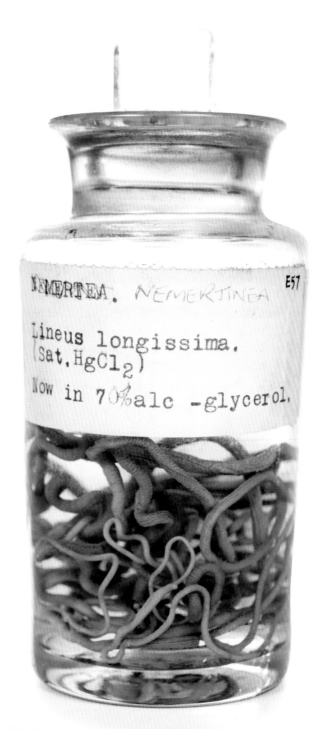

A preserved bootlace worm *Lineus longissimus*, the species that may be the longest animal in the world.

entangle them before drawing them in. Some have a stabbing needle in their proboscis which probably helps to convey toxins into the victim's body.

Nemerteans feed either by ingesting the whole animal or by sucking out its body contents through a hole it makes in their surface. Interestingly the proboscis isn't actually anatomically associated with the mouth or digestive tract: it is squeezed out of a fluid-filled sack running above the gut. The contraction of rings of muscles around the sack shoots the proboscis out like toothpaste out of a tube. A very long and stretchy muscle then pulls it back in. Those that live on land can also use their proboscis to pull themselves along.

Unlike the animals that we've explored so far in this book, ribbon worms have a true through gut with a mouth at one end and an anus at the other. It seems sensible to speculate that having a gut in which food passes all the way along the body in one direction would allow for the extraordinary body lengths seen in this group. It would be hard to imagine having to pass waste back from the closed end of a 50m-long gut back up to the mouth for expulsion.

They also have a circulatory system: they have blood vessels but no heart. Blood is pumped by the contraction of muscles around the body squeezing it along the vessels by constriction. This is more or less how our own veins work (our heart is only used to pump blood through arteries), but unlike in vertebrates, ribbon worm blood doesn't necessarily only flow in one direction.

Ribbon worms have a well-developed nervous system with clusters of coordinating nerve cells (ganglia) that form a brain, and longitudinal nerve cords. Together these control the muscles that run below their 'skin', and receive and interpret information from clusters of simple eye spots (which can tell light from dark but can't form images) and chemosensory organs in the species that have them.

Despite these organ systems adding significant levels of complexity to these worms' bodies, nemerteans can reproduce asexually by simply breaking into pieces (which can also happen through damage from would-be predators or by environmental action). The broken sections can then regenerate into new individuals. However the majority of nemertean reproduction is done sexually by breeding between two separate sexes, mostly by releasing eggs and sperm to meet in the outside world.

With regards to their evolutionary relationships, although they are superficially similar to flatworms they are not closely related. Their similarities probably arose because they have independently evolved similar features to live in similar niches, or because they have both retained ancient characteristics from a distant shared ancestor. Molecular and anatomical comparisons suggest nermerteans are more closely related to annelids and molluscs.

7 King Ragworm

PRESERVED SPECIMEN

Segmented Worms: The Annelids

Earthworms, bristle worms, leeches, peanut worms, beard worms and spoon worms all belong to the phylum Annelida. They are a diverse group of 15–20,000 species that run the full gamut of animal ecological roles. There are herbivores; predators; composters; soil-, grit-, and sand-eaters; those that derive nutrients from hydrothermal vents; scavengers; internal and external parasites and filter-feeders.

Describing annelids is quite difficult: across their diversity of forms and lifestyles they have no major unique characteristics that define them as a group. They are known as the 'segmented worms', but segmentation is a feature that arose in their ancestors and was inherited by them – as such it is not unique to them (and indeed many modern species have all but lost the appearance of segmented bodies). At best, annelids can be defined by a specific combination of features, each of which is shared with other groups.

Segmentation is a neat evolutionary trick whereby the same basic anatomical layout is repeated again and again down the length of the body. Each segment has its own set of organs (although some, such as the nerves, muscles and blood vessels run through the partitions between segments) – even its own local brain. This allows for diversification and specialisation, as different parts of the animal can take on new forms and functions without compromising the overall requirements of the animal. The rearmost section contains the anus, and the segment at the front is the head. Owing to the complex sensory stimuli processed by many annelids, they have a complex brain at the head end, along with sensory receptors.

Segmentation also helps fluid-filled worm-shaped animals to move. By working the muscles in one segment, fluid is squeezed into the next one. Coordinating such muscular constrictions allows the worm to push forward and also to burrow. It also provides some degree of insurance against damage – if a few segments are injured the rest of them can take up the slack without the whole animal significantly suffering.

Annelids have an incredible range of kinds of eye, including light/dark sensors, complex eyes with retinas and lenses, and even compound eyes comparable to insects. They have some other structures that are poorly understood but are believed to also react to light. Some of their eyes are found in the head, associated with the brain as in most eyed animals, but they have them in other places too, such as down the length of their trunk or on the structures that project out of their bodies.

Historically, most annelids were placed into one of three taxonomic groups: polychaetes, oligochaetes and leeches, but the boundaries between these groups have become a bit fuzzy. The polychaetes – or bristle worms – have very obviously segmented bodies covered in bristles (chaetae) made of the complex polysaccharide chitin (which also forms insect exoskeletons – their hard outer coverings), and each segment has extensions coming out sideways. These 'parapodia' (meaning 'side-feet') are important in locomotion: sometimes they are used as paddles for swimming, but most often polychaetes walk on them. Polychaetes can have several tentacles on their heads to assist with finding food, with eyes, structures for detecting gravity and chemosensory organs. They can sport serious-looking jaws, as in this specimen of the omnivorous king ragworm, and in other species these can be strengthened with metal compounds and also inject venom.

The oligochaetes, typified by the humble earthworm, have neither prominent sensory organs nor parapodia, and few chaetae. The raised structure near the head of an earthworm – the saddle or clitellum – is a feature that unites this group. It is used to make a mucus cocoon that detaches and closes up to protect growing embryos as they develop.

Finally, the group that contains the leeches (the Hirudinea, closely related to oligochaetes) deserves some attention, as its members have some incredible adaptations to their bloodsucking lifestyles.

There seems to be inherently more respect given to predators than parasites. When a velvet worm (which is not an annelid) spews strings of glue all over its prey, it is considered spectacular and impressive, earning airtime on high-budget natural history documentaries of the highest calibre (chapter 11). Compare that with the distain heaped upon leeches, who are at least decent enough not to kill their prey.

Most leeches have a sucker at each end of their body, to allow them to attach themselves firmly to another animal during each step of their caterpillar-like walk. It isn't precisely known how leeches detect their next victim, but terrestrial species could well be using vibrations, carbon dioxide or heat given off by vertebrate hosts. They attach themselves to something like a leaf and stretch out waiting to hitch on as something passes. However they do it, it pays to walk at the head of a group in leech-ridden habitats, as the stimuli given off by the group leader are likely to alert the leech to position itself to catch the people walking behind.

Once on board their host they find a secluded spot where they are not likely to be brushed off or spotted (like the groin or underarm) and get to work sawing a hole with their three serrated blade-like mouthparts. They inject chemicals to widen the blood vessels and to hold back clotting, as well as anaesthetics to avoid being noticed, and antibiotics.

Their blood meals, by their nature, already contain a lot of nutrients that can be absorbed directly by the leech, but leeches give a home to a community of symbiotic bacteria that digest other elements of the blood for them.

Leeches also show a wide variety of parental care behaviours, including carrying their eggs and young – even feeding them as they grow. This group of incredible parasites should be applauded rather than reviled (though I may withdraw that statement next time I'm back among them in the Australian rainforest).

Perhaps because of their abundance in the environment (which increases the odds of being preserved) or because of their collagen cuticles and chaetae, annelids have a better-established fossil record than other soft-bodied animals. Fossils confidently held to be annelids date back to the Cambrian period, 520 million years ago, but they almost certainly predate that.

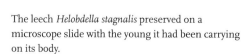

The leech *Helobdella stagnalis* preserved on a microscope slide with the young it had been carrying on its body.

8 King Scallop

TRAMOND WAX MODEL

Success With Shells And Suckers: The Molluscs

We live in a mollusc-rich world. To date, over 117,000 living species have been described within the phylum Mollusca. Gastropod molluscs, the group that includes snails, slugs and sea slugs, number an incredible 100,000 species. That's twice the number of vertebrates that have been described (and no doubt we've been focussing our efforts disproportionately on the vertebrates, so that discrepancy is likely to be even greater in reality).

Given that most molluscs have a hard shell, they have left us a rich fossil record to explore. Most of the groups alive today had arisen by 530 million years ago, in the Cambrian period, and there are even fossils from the enigmatic Ediacaran fauna, dating to 635 million years ago, that could be early members of Mollusca.

For much of their bewilderingly long history molluscs have been extremely diverse and abundant, just as they are today. While the Mesozoic era – comprising the Triassic, Jurassic and Cretaceous periods – is known as the 'age of reptiles' because of the impressive array of dinosaur, pterosaur, ichthyosaur and plesiosaur fossils of this age, there are those who roll their eyes at humanity's obsession with giant reptilian teeth and claws and insist this era should be known as the 'age of molluscs'.

At that time there were innumerable swimming shelled molluscs – the extinct ammonites and belemnites – as well as a dizzying diversity of molluscs from groups still living today. Molluscs even formed entire reef systems during the Cretaceous: the shells of rudist bivalves (also now extinct) could build reefs in seas that were too warm for corals to survive.

It's not just the Mesozoic that mollusc-fans think should be named in their honour. There are also those who insist the Ordovician period (485–443 million years ago), the Silurian period (443–419 million years ago) or even the whole of the last 500 million years should be dubbed the 'age of molluscs'. We now live in what's called the 'age of mammals' – the Cenozoic period – which is rather unfair on the molluscs, which outnumber mammal species by more than twenty to one (not to mention the more than 900,000 species of insects that have been described).

Modern molluscs include the gastropods, the bivalves (those with two parts to their shell like the scallop pictured on the previous page, oysters and mussels), the cephalopods (squids, octopuses and nautili, all with large eyes and complex brains), chitons (which resemble flat slugs protected by eight shielding plates) and tusk shells.

On the face of it there is not a lot that appears to unite these disparate groups, but the general molluscan body plan includes a tough integument or 'skin' called a mantle on their upper and lateral surfaces, and a single muscular foot that is used for locomotion. The mantle can secrete calcium carbonate crystals and a protein called conchiolin to bind them. In most

An octopus with its arms spread. Its mouth is in the centre with a sharp bird-like beak.

cases this forms a protective shell around the animal's organs, but in some groups the shell has either been lost or positioned internally. Most molluscs also have an unusual feeding apparatus called a radula, which is a ribbon-like structure covered in teeth made of the tough polymer chitin. The radula moves like a conveyor belt to rasp away at whatever the animal is eating.

It seems hard to believe that the slimy foot of a snail is the same anatomical structure as a cephalopod's arms and tentacles, but the front part of the foot has indeed been modified to form these predatory and sexual tools. They generally sport suckers and often vicious-looking teeth. A cephalopod pedant would note that most octopuses do not have tentacles, but instead there are eight arms. Squid also have eight arms, but in addition they also have two tentacles: arms are covered in suckers along their length; while tentacles only have them on pads near their ends. In any case, they are used to grab prey to bring to the mouth near their base (and sometimes deliver packages of sperm). Here a parrot-like scissor-action beak made of chitin rips the animal apart, the morsels of which are then drawn into the mouth by the barbed radula.

The diversity of molluscan lifestyles includes species that occupy most ecological roles, and occur in and near the sea, in freshwater and in nearly every terrestrial habitat. As well as the famous grazing gastropods, this group includes predators that drill their way into other shelled animals or chew up softer species (see the blue sea slug, chapter 58), and even parasites. The marine chitons are grazers too. Bivalves are generally filter-feeders, using their gills to trap tiny particles from the water, but shipworms are bivalves that bore into wood (infamously ships' hulls), using symbiotic gut bacteria to digest the material as they go. The cephalopods are all predatory.

The incredible variety of the molluscs is impressive in its own right, but the range of body forms, ecological niches and habitats that these animals occupy means they have had an extraordinary influence on the diversity of other animal groups. The development of the molluscan shell has forced the evolution of countless adaptations in predators across the animal kingdom which seek ways of getting to the meat: sharks, rays and many other fishes have evolved nutcracker-like crushing teeth; walruses have evolved the means to daintily suck their flesh out of the shell without breaking it (chapter 37); thrushes have developed incredible problem-solving tactics and crack their shells on favourite anvil stones; mantis shrimp have developed the anatomy to punch their way in (chapter 69). Not to mention every species of hermit crab, which rely on discarded gastropod shells for a home. To catch the cephalopods, animals as different as sperm whales, penguins and seals have evolved the super-fast hunting behaviours. In this way, diversity drives diversity, and molluscs are certainly diverse enough to have had a huge impact on life on earth.

9 Bryozoan

DRY SPECIMEN

Moss Animals: The Bryozoans

When I was at university studying zoology, many of my family asked if that meant that I was going to be a vet. The assumption was that anyone who was interested in animals must aspire to work with cats and dogs, as pets were their key frame of reference onto the animal kingdom. It just didn't occur to them that people could study or work professionally with non-domesticated animals from beyond the human world.

That may seem a little naïve but it's actually not far off an assumption that is probably held by most people: that zoologists would want to study charismatic animals. Some people would be surprised to learn that there are many zoologists out there studying entire animal phyla that most people haven't heard of.

Pentapora foliacea, the largest bryozoan to be found around British coasts. This specimen is about 30cm across.

My interest lies with mammals, which are certainly at the glamorous end of the spectrum – perhaps only surpassed by dinosaurs – but I have dear friends who are obsessed with flies, corals, squid and butterflies (and get paid for it). Although I think mammals are amazing, it's not at all hard to see the attraction of these other groups: they are globally important, highly visible, diverse creatures that live in incredible ways and exhibit some extraordinary behaviours. These people often get a bit riled about the amount of attention that cute and furry mammals are afforded.

Spare a thought, then, for the bryozoologists: the people who study bryozoans. I hope they will forgive me for saying that bryozoans (which means 'moss animals') are among the least charismatic groups of all. Also known as sea mats, they are miniscule, filter-feeding colonial animals – each less than a millimetre long – that are most commonly found growing on the surface of rocks, shells, fronds of seaweed and other marine surfaces. Few species can move from the spot where they first settle.

Although they may be tiny individually, the colonies can approach a metre wide. Some even form three-dimensional branching structures reminiscent of corals or sponges. Each individual has a little ring of tentacles – also somewhat coral-like – to trap passing plankton, but their bodies are quite a lot more complex than corals. For example, unlike corals they have an anus and a simple brain.

They are very common: if you've snorkelled, rock-pooled or even strolled along the beach you will have encountered bryozoans. Few people pay much attention to them, but if you can remember ever having seen a piece of seaweed with a fine, white net-like structure growing on it, that was one of the more noticeable bryozoans. They rarely enjoy national press.

Unlike corals, bryozoans don't really do reef-building, so one couldn't really call them environmental engineers. However they are rather diverse – around 5,500 living species (about the same as mammals) and around 15,000 from the fossil record, so they certainly do keep the few bryozoologists in the world busy.

One of the world's first bryozoologists was the Grant Museum's founder, Robert Edmond Grant, and it is *almost* arguable that his interest in bryozoans changed the world. As discussed in the sponges chapter, Grant studied extremely simple animals to see what they could teach him about the origins of the animal kingdom. In the 1820s bryozoans were referred to as 'zoophytes', which means 'plant-animals', and Grant was rather remarkable as even at that early stage he was seeking connections between the plant and animal kingdoms. Grant controversially believed that the simplest of animals should resemble the simplest of plants. He was of course correct.

Trained as a medic, but very much focussed on comparative anatomy at the University of Edinburgh, Dr Grant took the student Charles Darwin under his wing. Darwin was sent to Scotland to study medicine at the age of 16, but he found his excursions with Grant more engaging than his medical training. Grant would take Darwin collecting on the shores of the Firth of Forth, looking at the invertebrates they found there and enthusing about his interest in them.

In his autobiography Darwin describes a collecting trip where Grant 'burst forth', enthusiastically imparting his excitement for evolutionary thinking on the teenage Charles. In the role of mentor, Grant taught Darwin how to use a light microscope, and helped him make some novel observations on bryozoans: Grant had clearly rubbed off on him. Unfortunately there is a small scandal here – although Grant helped Darwin publicly present his ideas in what would have been Charles' first scientific contribution to society, Darwin never put the work to paper. So when Grant mentioned some of Darwin's discoveries in his own accounts of bryozoans without crediting him, there are some accusations of plagiarism.

Writing decades later, Darwin only begrudgingly concedes that his time with Grant – and specifically his outburst that day on the Firth of Forth – must have had an impact on the development of his own evolutionary ideas. In a way it's a shame as it comes across as Darwin being unwilling to share any credit with his old mentor in how he came to his world-changing ideas, but it could be because he was still annoyed at Grant for not mentioning his work on bryozoans.

While mammals, dinosaurs and even flies and corals get all the attention, sometimes the tiny, barely noticed underdogs like bryozoans have the potential to change the world. Indeed today they are being studied for their possible contributions to medicine: the unique species of bacteria that have evolved to live symbiotically inside bryozoans produce chemicals that may be of interest in the fight against cancer and Alzheimer's. Next time you meet a bryozoologist, please buy them a drink.

10 Brachiopod

FOSSIL SPECIMENS

Lamp Shells: The Brachiopods

Some animals are most commonly defined by what they are not. The first thing that most people say about horseshoe crabs, for example, is that they are not crabs (see chapter 63). Likewise flying lemurs are not lemurs, camel spiders are not spiders and golden moles are not moles (not to mention that flying lemurs can't fly, camel spiders don't eat camels (as was once thought) and it would be generous to describe the brown fur

The fossil brachiopod *Spirifer*, a diverse genus that lived from around 455 to 265 million years ago.

A modern brachiopod *Lingula sp.* washed up on a beach. Its long anchoring pedicle can be seen growing out from the shell.

of golden moles as golden). I feel rather sorry for these animals that are denied a unique description of their own in this way; their status as being 'not something else' is given as the most interesting thing about them. Brachiopods are such animals.

The first thing that is most often said about brachiopods is that they are not bivalve molluscs. Bivalves are the huge class of molluscs including mussels and clams that are characterised by having two shells (called valves) – one on the left and one on the right. Brachiopods are also small marine invertebrates with two shells, but they have one on the top and one on the bottom. Most bivalves are roughly symmetrical down the plane between their two valves: each shell is normally an approximate mirror image of the other. Brachiopods by contrast are roughly symmetrical down the plane that runs across the midpoint of their shells: their left side is an approximate mirror image of their right.

Brachiopods are not molluscs at all – they belong to an entire phylum of their own. A phylum is the largest of the main taxonomic subcategories: after 'Kingdom: Animal', it goes Phylum, Class, Order, Family, Genus, Species. Mollusca is one phylum, Brachiopoda is another. A body plan with a pair of opposing shells has evolved independently in the bivalve molluscs and the brachiopods.

Although the main levels of taxonomy are listed above, there are innumerable hierarchical steps in between these, because it's helpful to put species into more than seven nested groups, allowing taxonomists to provide a lot of information about how closely the species in question is related to other species (apes, monkeys and tarsiers, for example all belong to the infra-order Haplorhini – the primates with dry noses).

One of the biggest groupings of animals – at the super-phylum level – is the deuterostomes, which includes echinoderms (such as starfish and sea urchins) and chordates (which is mostly comprised of the vertebrates, including us). These very different groups are united by the way in which their embryos first develop a digestive tract. Brachiopods were long grouped with the deuterostomes and considered close relatives of the echinoderms and chordates. Groups that fall anywhere near the evolutionary pathways of humans are generally afforded more interest than more distant species, as humans are egotistical.

In recent years, however, comparison of the genetic and other molecular data across invertebrate groups has shown that brachiopods do not belong with the deuterostomes. Instead they fall into another massive grouping (the Lochotrophozoa) with ribbon worms, bryozoans, annelid worms and – as it would turn out – molluscs.

Brachiopods have a vast fossil record – they had appeared by the early Cambrian period, 540 million years ago. They were extremely common throughout the Palaeozoic era, even forming vast reefs. Their heyday appears to be behind them: over 12,000 fossil species have been recognised, but there are less than 400 species alive today. They survived the 'great dying' at the end of the Permian period, when 70 per cent of life on land became extinct, along with a staggering 96 per cent of marine species (see chapter 83). However, they began to decline after that. It is believed that they were on the whole outcompeted by a growing diversity of bivalve molluscs (though there are habitats today where brachiopods still dominate).

Brachiopods are filter feeders. Most species have an arm-like structure called a pedicle that extends out of a hole between the shells to anchor them into marine sediments. From there they filter-feed – extracting tiny organic particles and plankton out of the water for food. They crack open their shells just slightly, and most of the space inside is occupied by the apparatus for filtering. They have tentacles covered in tiny hairs (cilia) that waft water in, and they pass the particles down towards the mouth for sorting.

Interestingly not many animals feed on brachiopods – they seem to find them distasteful. Despite their huge abundance in the past (some sediments appear to be formed of little else except brachiopod fossils), there isn't much evidence of them forming part of extinct diets either. The occasional shell is found with the characteristic drill hole of a predatory snail, but it might be sensible to assume that these too had mistaken them for a bivalve.

11 Velvet Worms

Velvet Worms: The Onychophorans

On a zoological trip to Tasmania's temperate rainforests, eager to locate some of the world's most extraordinary animals, I went into a national park office and said to the ranger, 'I hear you have velvet worms.'

Clearly my meaning was mistaken as she visibly bristled, blushed and squared her shoulders at me exclaiming 'Velvet ... Worms ...!?' enunciating every syllable. 'I most certainly do not!'

I took a second to compute what had happened, before laughing in embarrassment. Rather than inappropriately accusing a stranger of having a parasitic worm infestation, I was in fact hoping for advice on how to find these enigmatic predators among the rainforest leaf litter. Needless to say the ranger didn't know anything about them.

As you have probably gathered from the diversity of 'worms' that have already been mentioned in this book, there isn't really a definition for what a worm is – biologically speaking, there is no such thing as a worm. Not all 'worms' are closely related to each other and some are even on very different branches of the evolutionary tree. It's not particularly nuanced, but most people would probably say that to be called a worm an animal would have to be relatively long and thin, and wouldn't have any legs. Velvet worms – the 100–200 members of the phylum Onychophora – don't really fit that description.

Their name means 'claw-bearers', giving away the fact that velvet worms do in fact have legs – up to 43 pairs of them. They are soft and unjointed little stubs. Without a hard exoskeleton (they are just covered in a thin chitin cuticle), velvet worms' legs and bodies are kept rigid by the pressure of their internal fluids. By moving this fluid around the body, and by using muscles, velvet worms use their legs to walk.

At the tips of their legs are little pairs of claws made of chitin, which allow them to gain a grip as they walk. It would seem that the head has developmentally absorbed some of these legs, as a pair of the walking claws has been modified in the mouth to form sharp, pincer-like jaws.

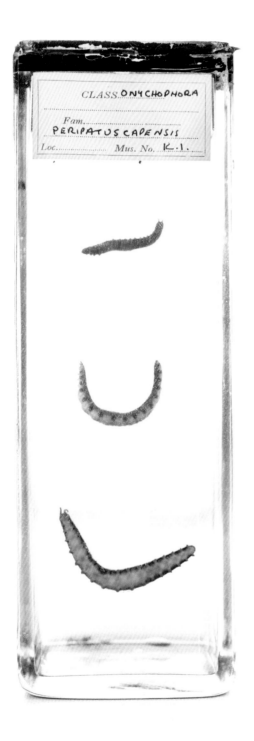

Preserved velvet worms *Peripatus capensis.*

Another pair of modified head-legs have evolved to discharge a remarkable predatory feature: velvet worms spray two jets of glue at a high force out of glands on their faces in order to trap prey. As the movement of the glands oscillates, the two jets effectively form a glue net, immobilising the invertebrates they feed on.

The sticky nature of the glue breaks down quickly in air, so they don't have long to get to work. Biting voraciously with their sharp mandibles, they inject a powerful saliva that starts to digest their victim while it is still alive. They then suck up the resulting soup.

Producing the amount of glue spray involved in such a manoeuvre is energetically very expensive, and so velvet worms have to judge whether or not to attack. They have small but complex eyes with a lens and a retina (evolved completely independently from other similar eyes in the animal kingdom), but it's not entirely clear how they are used in hunting. Velvet worms are extremely susceptible to drying out and so they only hunt at night, when it's cooler and more humid. This means that there is little light for their small eyes to see by, plus they are also positioned

A velvet worm firing the jets of glue it uses to catch its invertebrate prey.

behind two large antennae that would seem to obscure their forward-looking vision.

It is these antennae that are the key to their hunting skills. They use them to rapidly but very gently tap their potential prey to get a sense of size and shape, and they will only launch an attack once they have judged it to be big enough to justify the cost in glue, but small enough to stand a chance of being restrained (assuming it hasn't run away by then).

Of all the animal phyla alive today, velvet worms are the only one to be restricted entirely to a life on land. Their overall shape and appearance had led to a belief that they represented an intermediate evolutionary form between annelids and arthropods (the huge group that includes insects, arachnids, crabs and the like). We now know this isn't the case, as annelids are quite distantly related, but velvet worms do share a close ancestor with the arthropods. One of the key similarities between them is that both velvet worms and arthropods have to shed their outer coating in order to grow – the exoskeleton in the case of arthropods, and the cuticle in velvet worms. Both are formed of the polysaccharide chitin.

One remarkable thing about some species of velvet worm is that they show complex social behaviour that one might not think worthy of a worm, which just goes to support the contention that 'worm' is a pretty useless biological concept. These species show social behaviour more commonly associated with lions. They form tight-knit groups that hunt cooperatively to bring down larger prey, and then they feed together. As with lions there is a pecking order: no one gets to eat until the dominant female has had her fill.

12 Dragonfly

COMPLETE FOSSIL

The Most Successful Animals On Earth: The Arthropods

The dominant group on earth, whatever our egos might tell us, is unquestionably the arthropods. Insects, crustaceans, arachnids, centipedes, millipedes, trilobites, sea spiders and horseshoe crabs are all arthropods. Together they number significantly more species than all the other animal groups put together. So far around 1.2 million arthropod species have been described, but that number is likely to be a minute proportion of the reality – tens of millions of living arthropod species may yet to be described.

Arthropods represent 80 per cent of the world's known animal biodiversity, which is just one way of saying that for every five species on the planet only one of them isn't an arthropod. The world would be unrecognisable without them. Their role in nearly every ecosystem is extraordinary; since they form a huge bulk of edible biomass, a *lot* of animals and many plants, fungi and bacteria rely on consuming arthropods for food – and they are likely to form an increasing proportion of human diets as we seek more sustainable food sources. They also play a fundamental role as soil aerators, as composters, scavengers and waste-disposers, as gardeners, as grazers and browsers, as vectors of disease, as fertilisers, as parasites, as pollinators, and as predators. They are nearly everywhere and they do nearly everything.

Every taxonomic group is defined by a list of shared unifying features. However in many groups with high diversity (the molluscs, for example) the reality is that their evolutionary histories have modified these underlying features to such an extent that they are barely recognisable. It is therefore somewhat surprising that the most diverse group of all – the arthropods – has largely conserved their shared set of identifying characteristics. Most arthropods are easily recognisable as arthropods.

The word arthropod means 'jointed foot'; one of the traits that defines the group is that they have legs with joints in them. Their bodies are segmented and covered in a hard, thick exoskeleton made of chitin and sclerotin (a rigid protein) which must be shed and rebuilt in order for the animal to grow. Moulting the exoskeleton also allows them to redevelop body parts that have been lost or damaged. Arthropods have at least one pair of compound eyes (or in arachnids, simple eyes that appear to have evolved from compound eyes). Compound eyes are complex clusters of individual optical components, each with their own lens and light-receptors. It is a quite different solution to vertebrate vision, but at least as effective; indeed the comparison of images formed by neighbouring components gives the arthropods unrivalled ability to detect movement.

The exoskeleton provides arthropods with a good degree of protection against predators and water loss. Perhaps most importantly, it behaves like own our skeleton in providing rigidity and solid points for muscles to attach. With firm anchoring points, this allows muscles to grow long and strong. Combined with the sturdy scaffolding to support their weight, and

A 155-million-year-old fossil dragonfly *Protolindenia wittei*, from Solnhofen in Germany.

the defence against desiccating, this allowed arthropods to be the first animals to colonise the land, over 400 million years ago, and then to master it. Likewise rigidity and securely fixed muscles allowed one arthropod group – the insects – to take to the air.

Insects represent around three quarters of arthropod diversity – most animal species are insects. Their success is in no small part down to their capability for true flapping flight, which allowed them to occupy a whole new suite of niches. Insects were the first animals to evolve powered flight, around 350 million years ago. It is an incredibly valuable skill that has only arisen three other times in the history of life – in the birds, bats and extinct pterosaurs.

Given their diversity, it will come as no surprise that the arthropod lineage is an ancient one. The great diversity of animal life we see today really took off in the Cambrian period, when most of the modern lineages appeared relatively suddenly in the fossil record. Throughout this book, the probable 'start date' for very many of the groups described all fall within a short window around 540 million years ago – a period known as the Cambrian explosion (see chapter 20).

The period preceding the Cambrian is known as the Ediacaran (635–542 million years ago), and rocks dating to this age show a tantalising array of enigmatic fossils that palaeontologists have long struggled to ally with modern groups. Of course many of them may have gone extinct at the end of the Ediacaran, but at least some of them *must* have given rise to the animals we see today. Theories for which Ediacaran fossils relate to which modern groups abound – very few with any degree of certainty.

Those linked to the arthropods are no different – Ediacaran fossils from around 550 million years ago have been proposed as arthropods or arthropod relatives, but it remains extremely unclear. Definite arthropod fossils are found in the earliest Cambrian – the most familiar of which would be the trilobites (chapter 83). They disappeared some 270 million years later, which is an extraordinary reign of scuttling around on the sea floor. Trilobites were a significant presence in animal life for half of all the time that has passed since the Cambrian explosion.

Since their first appearance, to some approximation the story of animals is essentially the story of arthropods – surely the most significant animal agents in the history of time.

13 Roundworms

Keeping It Simple: The Nematodes

The number of individual animals alive on earth today is beyond comprehension. What is equally inconceivable is that four out of every five individuals among that number would be a nematode worm. Only 20 per cent of living animals today are NOT nematodes. In short, the world is home to a lot of nematodes. If I were to have selected objects for this book proportionately to their abundance, eighty of the one hundred chapters would have been about nematodes (so the one that they have actually been given seems a little unrepresentative). That number rises to more than 90 per cent when looking at animals on the sea floor. They are said to be so numerous that if everything on the planet that isn't a nematode were to become invisible (including all other living things, all water and all continents, mountains, rocks and earth), the world's features would still be vaguely recognisable from space because of the film of nematodes that coat them.

Not only are they astonishingly numerous, but they are incredibly diverse. To date nearly 25,000 species have been described. This is however a gross underestimate and may account for less than a quarter of a percent of their true diversity: nematode taxonomists have a long way to go.

They live in every habitat, in every water body and on every landmass. They are everywhere. Look at a drop of water from a puddle under a microscope (or even a magnifying glass) and you are likely to see nematodes.

Preserved specimens of the gut parasite *Ascaris sp.*

There are predatory nematodes and there are herbivorous nematodes. Pretty much every animal and plant is parasitised by nematodes. They are even found living in rock several kilometres below the earth's surface, feeding on bacteria that lived in the cracks. They are at the poles and in the forest. There are quite possibly some inside your body right now. Some species burrow between the cell walls of plants; and others still make their livings on rotting carcasses. Some eat other nematodes. Like I said, there are a lot of nematodes.

Nematodes can have complex life cycles and most have distinct sexes. Some species produce eggs that can withstand extremes of temperature and aridity, waiting dormant for conditions to suit them before hatching. Some give birth to live young after the eggs develop inside the female's body. Some parasitic forms need more than one host to complete their life cycle: as a juvenile they may live in the muscle of one host animal, and wait for that animal to be eaten by a predator or scavenger before they can become adults. Once 'awakened' in the guts of the animal that ate their first host, they become adults, pair up and mate inside their second host's intestines. Once born, the next generation of juveniles then spread through their new host's body, and burrow into the muscles where they wait for the circle of life to turn again: they need their host to be eaten in order to complete the cycle. The common human parasite *Trichinella spiralis* lives in this way.

They come in a variety of sizes, more commonly less than a millimetre than more than a metre, but the biggest reaches 9m long (aptly a parasite on whales). Whatever scale they are living at, nematodes are all long and thin, with a recognisable head end. Their overall body plan is extremely simple: they have an outer surface comprised of a stiff cuticle containing layers of collagen covering their epidermis (the outer layer of cells). Below that they have just four muscle strips running the length of the body, and all movement results from the contraction of these muscles against the stiffness of the cuticle. They have a brain with a few nerves running down the body, into which the muscles project. This is the opposite of the system in every other kind of animal, in which the nerves extend into the muscles. Within all of that there is a tubular intestine (but no stomach) which connects the muscular mouth to the anus near the tip of the tail. They have no lungs and no heart (they are so thin that respiratory gases can simply diffuse through the body wall).

Considering the apparent simplicity of their anatomy, one might have been tempted to think that there aren't many ways in which evolution could modify nematodes to create much diversity. It is clear, however, the opposite is true. As I mentioned they may be the most diverse group on earth; it would seem that rather than gaining anatomical complexity, keeping it simple is an extraordinarily good route to evolutionary success.

14 Penis Worm

Penis Worms: The Priapulids

Do some animals look too boring to be in a museum? This is a serious question. There is an underlying struggle in museum displays to fulfil two sets of needs, and they have to do both to be successful: 1. To engage the visitors' interests, desires or questions that are sparked by their own experience of a topic; and 2. For the museum to tell the stories that it has identified as the stories it exists to tell.

The struggle comes when a display meets one of these needs but not the other. This issue is the same in the worlds of politics and media – do we tell the people what they want to hear, or do we tell them what we want them to know?

In natural history museums, we know that people like big animals, for example. Dinosaurs meet both needs above – people want to see them, and museums want to engage people in stories about them.

Worms, on the other hand, do not follow this formula. They are long and thin but beyond that regularly lack much in the way of visible features, or even a

A preserved penis worm *Priapulus sp.*

noticeable head. Worm-shaped animals don't really grab the attention of museum visitors, and museums are visual places.

However, although they look boring, there are endless interesting stories about the many groups that we call worms. A third of the animal phyla – the major taxonomic groupings – have common names including the word 'worm'. But can you think of many museum displays given over to worms?

The interesting stories of worms – their diversity, evolution, and the role they play in ecosystems – completely fulfil Need Number Two above: museums would love to tell these stories. The idea that there are so many worms, and that they are everywhere, could potentially spark a visitor to see the animal kingdom differently. Even the idea that taxonomically speaking there is no such thing as a worm might give people pause.

But ... is there any point in a museum display that nobody looks at?

The central concept of the story is that worm-shaped animals are – despite appearances – really interesting. The problem is, however, that unless people already know what they are, they look so boring that nobody stops to read about them because they don't fulfil Need Number One. Long, thin featureless animals don't engage people.

The failure at Need Number One means that Need Number Two is irrelevant. What's the point of displaying it if we can't get visitors to engage in their story? A populist approach to museum management would result in worms not going on display. Arguably this is what has happened in many or most museums.

As wrong as it sounds, museums make decisions like this all the time. Vertebrates make up just 2–3 per cent of known species, but invertebrates probably make up a similar proportion of museum display space. Natural history museums are deeply unrepresentative of nature – full of dinosaurs, whales and elephants, normally giving bugs and worms the occasional mini-display. There are exceptions, where museums have created stunning invertebrate-led exhibits to capture visitors' imagination (see chapter 93) but they are always dwarfed by the space given to vertebrates. This is completely understandable – museums need to display things that people want to look at. But is it museums' responsibility to be comprehensive?

At the Grant Museum of Zoology, where I work, there are actually quite a lot of displays including worms. This may sound taxonomically more ethical, but the truth is that hardly anybody looks at them. Except one, and it's this one: the penis worm. We regularly hear one visitor urgently whispering to another 'Look! A penis worm!'

Penis worms do not look spectacular, and on their own I doubt they would attract any more attention than the next 'worm', however, they have the benefit of being labelled, in big letters 'PENIS WORM', which for obvious reasons manages to catch the eye.

Penis worms belong to the phylum Priapulida, which is Latin for 'little penis'. Zoologists often use descriptive names, and there is no escaping the fact that these burrowing marine animals do resemble human penises. I very much doubt that the taxonomist responsible would have realised that they would be doing priapulids – and museums – a very big favour in selecting this common name.

Penis worms have a mouth on a swollen proboscis that is surrounded by recurved spines. As muscles push body fluid in and out of the proboscis making it periodically engorge and deflate, the mouth is turned inside out and the spines scrape and catch on whatever is in front of them. Penis worms use this as a means of burrowing into the sediment, and also engulfing food. Little is known about the biology of the 15–20 species in the priapulid group, but some are believed to feed on other kinds of worm, while others behave like earthworms and eat sediment to digest the organic material between the grains. Together with the preceding three groups (velvet worms, arthropods and nematodes) they belong to a group that is united by the fact that they need to shed their cuticle to grow – the Ecdysozoa.

Ironically, then, there isn't actually much to say about penis worms, despite the fact that they are the most talked about worms in the Grant Museum of Zoology.

15 Sea Potato

PRESERVED SPECIMEN

There Are Five Sides To Every Story: The Echinoderms

It is very common to see the evolutionary history of animals described in terms of increasing complexity, from extremely simple animals like sponges at the bottom of a ladder, to mammals at the top. The idea that humans are the pinnacle of evolution, with every other group being a progressive step in complexity on the way to the appearance of humans (or mammals) is incredibly pervasive. Looking at the tree of life, it is of course utter nonsense.

Biologically speaking, other than those groups that have a very low level of anatomical complexity, such as sponges, comb jellies and cnidarians (the ones that are not included in the Bilataria), it's hard to make a coherent argument that anything else is more or less complex. There are gradations of how elaborately a group has developed a certain feature (for example the vertebrates have done some reasonably impressive things with the brain), but in a sense these are just modifications on a theme. Thinking about progressions of complexity is also problematic as very often in evolution features that may be considered complex are *lost* as lineages arise that have no need for all the traits inherited from their ancestors.

At this point in the book we make a departure – apart from those three non-bilatarian groups everything in this section that precedes this point is equally closely related to us: the noble flatworm is no more distant a cousin than the honey bee. They all belong to the huge taxonomic division called the protostomes. The remaining phyla in this 'Understanding Diversity' section of the book (which together make up the deuterostomes), however, are included here in increasing degrees of relatedness to us. One of the deepest splits in the trunk of the tree of life is between these two giant groups, and animals on either side of the divide are no more or less 'complex' than each other.

One might assume that being highly mobile and displaying complex social behaviours – like bees and ants for example – indicates that a group is allied with us vertebrates, but these characteristics have evolved more than once and aren't always useful in thinking about our place in nature. Bees and ants are very distantly related to vertebrates. Instead, of the animals traditionally referred to as 'invertebrates' one of our most closely related groups is the echinoderms – the phylum that includes starfish, brittlestars, sea urchins, sea cucumbers, feather stars and sea lilies. These are neither highly mobile nor socially complex.

The slug-like sea cucumbers, plant-like sea lilies, pincushion-like urchins, biscuit-like sand dollars, tumble-weed-like basket stars and I-don't-know-what-like starfish do not appear to have a lot in common. However they are united by the presence of rigid calcium carbonate plates as a skeleton, bizarre worm-like tube feet and five-sided radial symmetry. The latter means that their bodies are divided into roughly equal fifths, so if they are rotated around one fifth of a circle, they will look more or less the same. Despite this, echinoderms are still members of the Bilataria, it's just that they've independently developed radial symmetry: their larvae are bilaterally symmetrical, demonstrating their evolutionary roots.

Radial symmetry is a sensible evolutionary solution to filter-feeding or suspension-feeding, as food could come from any direction (it is also seen in the cnidarians, for example), and suspension-feeding is what sea lilies, feather stars and many brittlestars do, by waving their long thin arms to catch passing debris.

Tube feet are a unique echinoderm feature. They are highly flexible and extendible little water-filled protuberances that stick out of the body and are used in both feeding and walking. Most have little suckers on the end, and each of them moves like a tiny leech on the prowl, waving in the water for something to stick to. Working in harmony hundreds of these tube feet creep the animal along the sea bed.

They can also be used in the bizarre reproductive and feeding techniques of starfish. Echinoderms have incredible regenerative powers: not only are their bodies able regrow lost limbs, but their limbs are able to regrow lost bodies. Although they do use sexual reproduction by releasing sex cells into the water to meet and fuse to form new embryos, they can also reproduce asexually by splitting in two. On a solid surface, tube feet on either side of a starfish's body can walk in opposite directions, until the animal tears itself in half. Likewise a single leg can walk away by ripping itself off to start life on its own by re-growing a new body, as can be seen in the specimen here.

Assisting the predatory behaviours of starfish, the tube feet are used to help pull open bivalve molluscs. After locking the solid plates of their body together around a bivalve, the tube feet force open a crack between the two parts of the bivalve shell. The starfish then pushes its stomach out of its mouth so that it is inside-out inside the bivalve, and releases its digestive juices before absorbing the resultant soup.

Tube feet protrude out of little pores on the shell of an urchin and are long enough to extend beyond the long protective spines so they can walk up vertical surfaces (this solves the problem of how flat-bottomed solid animals climb steps). It's a crucial adaptation to help them climb up the kelp on which many species graze. Incredibly, urchin spines are formed of a single crystal of calcium carbonate and are moved on a ball and socket joint on the shell.

Sand dollars and sea potatoes, like the one pictured on p. 55, are modified sea urchins adapted to burrowing. They use their spines and tube feet to push their bodies through the sediment, which they eat to digest organic particles between the sand grains. Their bodies have evolved to become bilaterally symmetrical (they are narrower than they are long) which allows them to push more easily through the mud, but the five divisions, showing their radially symmetrical origins, are still clear.

16 Graptolites

Acorn Worms, Sea Angels And Graptolites:
The Hemichordates

Positioned as either the closest relatives of echinoderms or one step closer than them to the group that includes vertebrates, the hemichordates are our close kin. This could serve as another humbling lesson in what our relations tell us about our place in the animal kingdom.

Living hemichordates fall into two rather different groups, neither of which could be described as spectacular. The acorn worms resemble saggy, wrinkly versions of earthworms, and live rather similarly to them, but in marine sediments. The pterobranchs or sea angels are small tube-dwelling colonial filter-feeders with feather-like tentacles. Together they

British graptolites *Diplograptus sp.* from around 450 million years ago.

only number about 100 living species. Despite their relatedness to vertebrates, these worm-like animals are definitely among those that museums seem to think look too boring to display.

At the Grant Museum of Zoology we have the honour of having described the pterobranch class: in 1877 the museum's curator E. Ray Lankester brought the group into taxonomic existence. This is really relatively late for a branch of this level to be first described on the tree of life. Still today little is known about living hemichordates in general.

In contrast to living hemichordates, there is a group of animals known only from the fossil record that displays a huge degree of diversity and survived from over 500 to 315 million years ago. They were the graptolites. For a very long time no one knew what graptolites were – they were a complete enigma. Their name means 'writing in stone', because they resemble pencil marks on slate. Graptolites were extremely common but not at all understood.

Although no one could say what kind of animal a graptolite was, or even what part of it had been fossilised, it didn't stop palaeontologists describing many different orders, families and species of them. Although they shed very little light on their own existence, the fact that they are highly abundant fossils which show rapid rates of evolution means that they are extremely useful for dating rocks. If another fossil were found alongside a certain species of graptolite, it is possible to use the latter to give an age of the former because graptolites vary a lot through time. They are called index fossils.

In the 1970s discoveries were made that showed that graptolites are in fact hemichordates, most closely related to the pterobranchs. Moreover they may be so closely related that pterobranchs could actually be surviving graptolites, which would make for an extraordinary comeback from the fossil record.

Hemichordates today are clearly a mere shadow of their former diversity. In the Palaeozoic, 400–500 million years ago they were the dominant animal plankton in the sea. Unlike modern pterobranchs, which live on the sea floor, many graptolites floated on ocean currents. Both groups are colonial with individual tentacled zooids poking out of little tubes in a cluster (though the structures formed by graptolites appear to have been more regular). Looking in today's oceans there really isn't a comparable group to the graptolites, ecologically speaking.

Too few fossils which indicate the soft body parts of graptolites have been found to know what they really looked like, but pterobranchs and acorn worms are united by a shared overall body plan. They are soft and cylindrical (and extremely fragile when handled, but that isn't too much of a problem for them as they share the echinoderms' ability to regenerate), and are divided into three sections.

At the front end is a conical proboscis (apparently acorn-shaped, which gives the acorn worms their name, but if I'm honest I can't see the resemblance), which gathers food. In acorn worms it is used to burrow and pterobranchs use it to climb up their tubes. Then follows a collar, which is where the mouth is and where the tentacles sprout from in pterobranchs. There is a stiff structure in the collar that resembles the notochord, a rod which supports the embryonic development of vertebrates and their relatives. It's unclear whether the rods are actually the same structure in the two groups, which would add to the evidence that they are related. The third body section is the trunk, which makes up the bulk of the animal (this can be considerable – acorn worms can reach 1.5m long). The gills that are found on the trunk also show affinities to the vertebrate guild.

It may be a mud-dwelling, uninspiring, saggy worm, but I quite like that I can look at a hemichordate and call it 'cousin'. It helps give a sense of perspective.

Pterobranchs (or sea angels) use their frilly appendages to filter particles from the water. They are graptolites' closest living relatives.

17 Lancelets

What Is A Fish? The Cephalochordates

Is a cod a fish? Yes. Is a shark a fish? Yes. Is a hagfish a fish? Yes (from the name it had better be, but then starfish aren't fish). Is a whale a fish?

School children everywhere would tell you quite clearly that whales are not fishes. They are mammals. But what makes a fish a fish? The problem is that if you can't define a fish, you can't say what isn't a fish. By logically following the rules of taxonomy, whales are actually fishes. Crocodiles are fishes. Birds are fishes. Frogs are fishes. People are fishes. Bear with me ...

Taxonomic groups are defined by unique *derived* characteristics. A derived characteristic is a feature that evolved in that group, and wasn't present in the group's ancestors. Derived characteristics of mammals include having fur and suckling their young – features absent in the animals they evolved from.

Rules of taxonomy also dictate that every group has to include, by definition, all of the groups that evolve from it. For example, primates are mammals because they evolved from mammals. Therefore we could not have a definition of mammals that excluded primates. Just because primates evolved to have the derived characteristics that define them as a group doesn't mean that they stopped being mammals. Also, just because apes have modified or lost some of the features seen in other primates – apes have no tail, for example – doesn't mean that they stopped being primates.

Put another way, every branch on the tree of life is considered to be a member of all its parent branches.

So what is a fish? There can be no definition of fishes that does not include everything that evolved from fishes. Mammals evolved from animals that evolved from fishes, and therefore so did whales. And people. Whales and people are fishes.

It is not possible to come up with a list of unique derived characteristics for fishes that excludes everything that evolved from them, but when in evolutionary history do we start counting everything as a fish? Some of the features we think of as fishy were present in their ancestors.

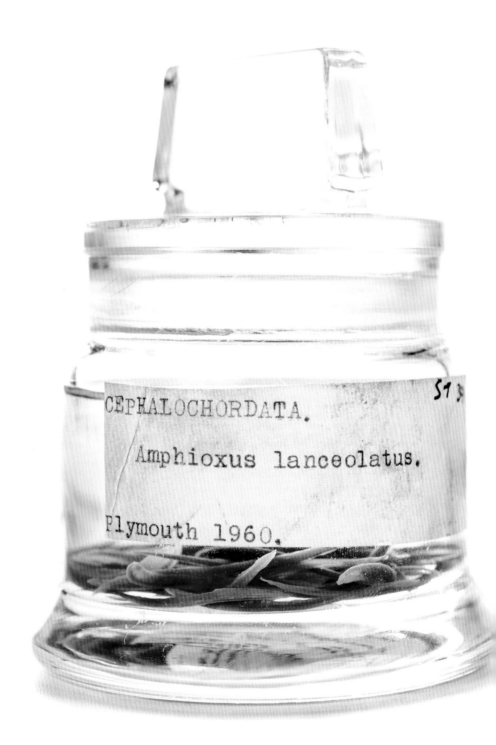

Most people would come up with a list of traits that were present at the dawn of vertebrates: fishes have gills, a tail and a backbone. This, therefore means that every vertebrate is a fish. (Remember, just because we don't have gills doesn't mean we're not fishes – gill slits appear and disappear in our embryological development.) But most people would be happy to include hagfish as fishes, and they don't have a backbone. So we need to push the definition of fishes even further back to include hagfish.

Can the fish net be spread any wider (no pun intended)? The next evolutionary step outwards away from hagfish (and the rest of the vertebrates) would be the sea squirts, which follow in the next chapter. These gelatinous blobs don't really seem like fishes. However a group that branches off even earlier on the evolutionary tree do seem pretty fishy: the animals pictured here – the lancelets.

They can swim, they have gills, a tail and are long and thin and are altogether fish-like. But they don't have a backbone. Instead they have a rigid but flexible rod called a notochord that runs the length of their body and is the precursor to the vertebrate spine. They also have a nerve chord that runs above the notochord along their 'backs' – just like us. By contrast the 'invertebrates' explored earlier in this section have nerve chords that run along their bellies.

The lancelets typify the animal group called the cephalochordates. Despite being rather unprepossessing eel-like filter-feeding animals which live half buried in shallow marine sediments, they have attracted extraordinary levels of interest (unlike penis worms, which do more or less the same thing). This is because humans are inherently self-obsessed, and lancelets are likely to be extremely similar to the very first members of the group from which we vertebrates arose.

This group – the chordates – contains the cephalochordates, the sea squirts and the vertebrates. The unique derived characteristic uniting this group is that notochord, the structure that acts as a useful anchor for muscle attachment along the body, and provides a stiffening function. Lancelets have retained this notochord very clearly in their semi-transparent bodies, and overall seem to have changed little from what we believe the earliest chordates looked like. This makes them interesting, as they help us understand our place in the world, and where we came from.

One of the earliest known chordates from the fossil record is the cephalochordate *Pikaia* from the Burgess Shale a little over 500 million years ago. This means that our branch of the animal tree forked off very early on, at a similar time to nearly all the other key groups. *Pikaia* appears very similar to modern lancelets.

Given that cephalochordates are so fishy, are they actually fish? Well, what is a fish anyway?

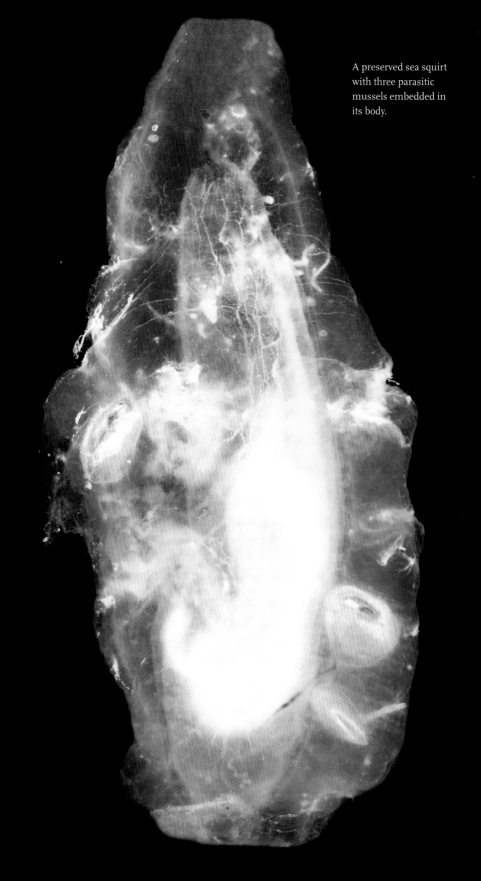

A preserved sea squirt with three parasitic mussels embedded in its body.

18 Sea Squirts

PRESERVED SPECIMEN

You Can't Choose Your Family: The Tunicates

The German naturalist Ernst Haeckel (1834–1919) promoted an idea called recapitulation theory, which stated that as an embryo developed it had to pass through successive stages that each resembled one of the species' evolutionary ancestors. As it grew, it would appear to relive its own evolutionary history. Most famously, this was used to explain why human embryos at one point have gill slits, as they were passing through a fish phase.

Haeckel didn't get it right – this is not what happens. The reason that developing humans have gill slits is that they share a common ancestor with animals with gill slits, and the genes and structures are still present for them. Although Haeckel is universally discredited, comparative embryology is a really important tool for evolutionary biologists. Studying how a species' features develop in the embryo can demonstrate that the feature in question is the equivalent structure to something in a completely different species. For example the wings of bats and the fins of whales go through the same embryological processes, developing from the same parts of the embryo, and controlled by the same genes, because they are differently evolved versions of the same mammalian arm.

It was from looking at the embryological development of a strange animal that zoologists were able to learn some surprising things about our own evolutionary affinities. Some of our closer relatives don't look a lot like us: they are the sea squirts.

You can't choose your family. This adage is undeniable when it comes to talking about our evolutionary history – we cannot choose to become unrelated to certain groups of animals. A sea squirt is effectively a tough, fluid-filled, translucent bag sitting on the bottom of the sea, spending its time sucking in water and feeding on the microscopic particles it finds there.

Most members of the sea squirt's group – the tunicates, or Tunicata – spend their adult lives permanently attached to rocks and other hard underwater surfaces processing gigantic volumes of water, sucking it in one hole (called a siphon), passing it through their gut lined with particle-filtering gills, and blowing it out another siphon. Despite looking like a wet paper bag, they actually have a brain, a heart and a kidney-like excretory system.

Sea squirts are not our direct ancestors, but they are very much on our branch of the tree of life. In recent years molecular studies have shown that the Tunicata are actually the closest relatives to our vertebrate lineage (until then they were considered the second closest relatives, after the eel-like cephalochordates (see previous chapter). Fossils suggest that the two evolutionary trajectories began their separate journeys over 500 million years ago. While that means we are not fantastically close relatives, they are more closely related to us than bees, octopuses, worms and every other kind of animal that isn't a vertebrate. This may be surprising because they look and behave quite a lot like sponges, the simplest of animals (which is what they were at one time considered to be).

For a very long time before the recent genetic studies, cephalochordates were believed to be the closest relatives of the vertebrates, basically because they look pretty fishy, and after all, the first vertebrates were fishes.

Although adult sea squirts look nothing like us or any members of the vertebrates it has long been known that they are our close relatives, and that comes from looking at their larvae. This is where comparative embryology comes in. The tadpole-like tunicate larvae have a rigid but flexible rod of tissue running down their back called a notochord. It provides an anchor for the body's muscles and – crucially – is the precursor to the vertebrate spine. This feature is also seen in cephalochordates and together the vertebrates, tunicates and cephalochordates make up the chordates – animals with notochords. It is the major taxonomic group to which we belong.

Above the notochord runs a dorsal nerve chord. In all other invertebrates with a central nervous system (except starfish and their relatives, which are also on our branch of the tree of life), the central nerve chord runs along the belly (a ventral nerve chord).

Although the 'official' name for the sea squirt group is Tunicata, they are most commonly known by the name Urochordata, which was given to them by the Grant Museum's curator E. Ray Lankester in 1877. While it is very much in popular use (among biologists), it is technically illegal according to the strict but fair laws of the International Commission on Zoological Nomenclature – who set the standards for naming animals – because someone came up with Tunicata first (it was the biological giant, and pre-Darwinian evolutionary forerunner, Jean-Baptiste Lamarck, in 1816).

19 Chimpanzee

MOUNTED SKELETON OF JUVENILE

Vertebrates: Animals Like Us

For the preceding seventeen objects we have encountered around half of the main branches of the animal family tree. On those branches, however, are the vast majority of animals alive today – the half that have not been included represent a mere sliver of animal diversity (most of which are exploring different ways to be a 'worm').

We have now reached the point where the object in question represents the group to which we belong: the vertebrates. Vertebrates are unquestionably extraordinary and well deserving of our attention – and not just because they are all animals like us. That said, I am somewhat loath to end this taxonomic section with vertebrates, as it has the potential for perpetuating a progressive view of nature – that everything that has happened in evolution was building up to the appearance of us.

That is not my intention, we are just one branch on the tree of life – albeit the only one capable of contemplating the concepts of taxonomy, evolution and our relative place in the world (as far as we know). Humans are not the pinnacle of evolution. We are the very best at being human, but sea slugs are the very best at being sea slugs and crabs are the very best at being crabs. None is 'more evolved' than the other.

Talking about taxonomy in an order that ends with vertebrates is commonplace in biology, even though the progressive view of evolution is utterly dismissed. Perhaps it is a hangover from earlier, less enlightened times. It can be justified to some degree because we are looking at the sequence that evolutionary branches sprouted off from their ancestral trunks. On our part of the tree (the deuterostomes), the echinoderms were the first branch-off, then the hemichordates, then the cephalochordates, then the tunicates and the vertebrates. In that sense we were the last to appear.

What that ignores, however, is that ours is not the only part of the tree: the protostomes – which are all the other 'invertebrates' except sponges, comb jellies, cnidarians and placozoans – have been going down their own evolutionary path completely separate from us. I could have talked about deuterostomes first, and then finished this chapter on any number of protostome groups, as many of them appeared at very similar times. The main reason I wanted to end on vertebrates is not because they are the most advanced group, but because the following section of the book will pick up with exploring some of the ways the vertebrates have diversified.

The vertebrates are masters of diversity: not in terms of pure numbers – there are far more kinds of arthropod and mollusc – but in terms of scale

and form. Like the arthropods, they have evolved to be adapted to more or less every niche (though there are very few vertebrates which are internal parasites). However, they have done it with a range of body sizes that is truly astounding. Even just among the mammals, the smallest is around 150 million times lighter than the largest: 150 million Savi's pygmy shrews weigh the same as a single blue whale. And there are many other vertebrates that are smaller than the smallest mammal.

Vertebrates have evolved incredibly intricate nervous, sensory, respiratory, excretory and circulatory systems to deal with a huge range of environmental conditions and lifestyles, and it is amazing that these shared physiological systems function over such a wide range of sizes.

All vertebrates have a brain protected by either a cartilaginous or a bony skull. Inconveniently for people who care about logical naming, not all vertebrates have vertebrae. It used to be thought that the living jawless fishes – hagfish and lampreys – sat outside the group as the closest relatives to vertebrates, (with all three together making up the group Craniata), but current thinking is that they all sit within it, and the terms 'craniate' and 'vertebrate' actually represent exactly the same grouping. As Vertebrata was coined before Craniata, we have to stick with vertebrate even if it causes confusion.

The vertebrates comprise the jawless fishes; the sharks, rays and relatives; the ray-finned fishes; the lungfish; the coelacanths; the amphibians; reptiles; birds and mammals. While the living jawless fishes and the sharks and rays have a cartilaginous skeleton, the remainder have a bony one.

Bone was an incredibly important adaptation, not least because it gave our ancestors the structural support to be able to bear their own weight out of water and follow the arthropods to exploit the benefits of living on land.

Weight-bearing was not its initial purpose, of course, as it first appeared in jawless fishes which were entirely aquatic. There was a whole suite of heavily armoured fishes that used bone as an outer covering, providing them with a significant advantage against predators. The earliest known vertebrate fossils are from Yunnan Province in China, dating to around 520 million years ago, which appear to show cartilaginous vertebrae protecting the notochord.

Bone proved a useful innovation – over 60,000 living species have it – and there are a number of theories for why bone first evolved. It provides solid sites for muscles to attach, it conveys protection to soft tissues below it, and a bony coating provides a degree of defence against water loss. It could have initially resulted from a need for animals to dump waste calcium from their systems. Conversely having bone provides a handy storage area for muscles to access the calcium they need to function. These are all functions that bone provides the animals that have it today.

PEDIGREE OF MAN.

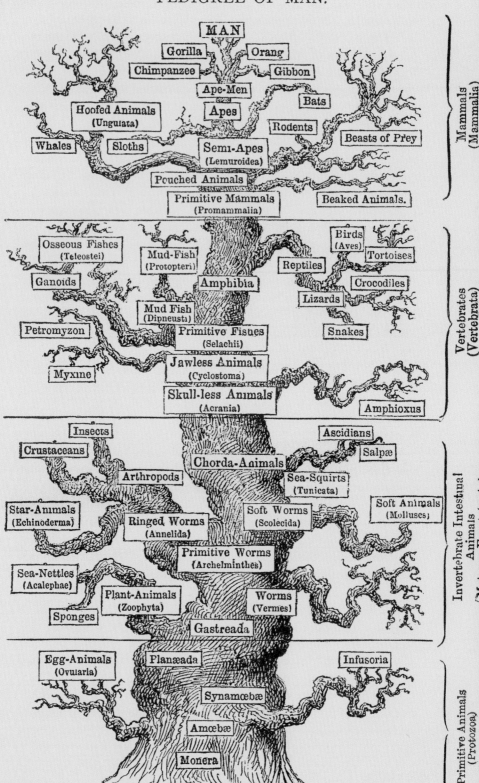

Part 2

Life's Turning Points

The reason the structure of a tree is such a well-used analogy for the history of life is that it makes a lot of sense. There is only one trunk – as life has only evolved once. Every so often something new happens, and a major branch forms. Each of those branches then has to grow on its own path, to an extent shaped by its neighbours, but also separate from them. The branches split to form smaller branches, and in turn they form twigs as life develops different features and new evolutionary trajectories are followed. Species are like the leaves.

So if each major branch indicates a large taxonomic group, the order in which the branches and sub-branches stem off from each other lets us know how the groups are related and how long they have been around. The junction where two branches go their separate ways is their last common ancestor.

What can be lost from the simpler tree diagrams is the relative size that some of the branches grow to: the arthropod branch, for example, will have hundreds of times the number of leaves of the mammal branch, and subdivide many times along the way.

Starting at the base of the trunk, we can follow a route up the tree, down countless branches as they split off from each other, until we get to a leaf. In so doing we have traced the evolutionary history of one species. The early parts of that journey will be shared with many species, but increasingly fewer as we travel towards the tips of the twigs.

In this section of the book, I have chosen ten objects that take us along the journey through the tree until it leads to the leaves that represent modern mammal diversity. It will take us along the branches, starting

Left: The 'tree of life' as it was understood in 1879, appearing in Ernst Haeckel's *The Evolution of Man*. This egotistical version places humans at the top of the tree, suggesting that evolution is progressive. In fact, we are just a single twig among millions of other species.

where all the major branches suddenly split off from the main trunk and begin the separate journeys of the modern animal phyla discussed in the previous section. We will travel down some of the length of the various branches representing fishes, until they split to give rise to the tetrapods. From there the appearance of life on land creates a new branch, and eventually mammals split off of that.

It's important to remember that this isn't *the* story of the evolution of life on earth, it's just one story: the mammals' story. We could follow other paths through the tree and discover how starfish came to be, or chickens, or beetles.

These ten objects represent ten turning points in evolution without which we wouldn't be here. In talking about evolutionary histories like this it is very easy to fall into tautological language. We could say things like 'in order to conquer the land, animals evolved weight-bearing limbs'. The problem with this kind of statement is that evolution has no direction and no purpose. It does not do anything *in order to* do anything else.

Changes occur through random mutation, and natural selection exerts a pressure that either encourages the spread of that mutation, or its removal from the population. Over time, this means that animals can move into new niches as changes are favoured by evolution that make it possible for them to do so. The animals were not *trying* to evolve, they just do.

Evolution is not progressive, so it is important not to think of these turning points as steps *up* the evolutionary tree. Moving away from the tree analogy, we should think of evolution as a flat, forking path, rather than a ladder: the 'steps' are paces, not stairs. Evolution did not start out with the 'intention' of one day creating mammals, and none of these turning points happened in order that we appeared. However, we would not have done so if they hadn't. The history of mammals, and every other group, is just an accumulation of random changes in anatomy and physiology. Our story is not 'better' than the wasps' story, for example (but it is better preserved in the fossil record). However most of us, I suspect, are more interested in our own story. This is the story of our branch, starting at the trunk.

20 *Ottoia* the Penis Worm

COMPLETE FOSSIL

The Cambrian Explosion: Where We All Came From?

Because of the shared cellular processes and structures, and the molecules involved, it is assumed that all life on earth only evolved once. All plants, fungi, bacteria, algae, animals and every other kind of known being share a common ancestor. It's very hard to put a date on the origin of life, but the oldest known fossils – described in 2017 – are at least 3.77 billion years old. These are apparently bacteria from rocks that formed in deep-sea vents in what is now Quebec, Canada and they may actually be as old as

4.3 billion years. This is remarkable as the earth only formed 4.5 billion years ago, and already these fossils are far more complex than what we expect the first life-forms to have looked like.

For a very long time living organisms were exclusively single-celled, like bacteria are today. Life became slightly more complex when eukaryotes evolved (cells which have nuclei to contain most of their genetic information, as well as other discrete organelles like mitochondria). The oldest known eukaryotic cells appear in the fossil record just under 2 billion years ago but they too remained single-celled for aeons.

Again, it is unclear what constitute the first animal fossils, but rocks of Ediacaran age, first discovered in South Australia and dating back to 630 million years ago, contain some very interesting organisms. They could be animals, plants, algae or something else.

While various attempts have been made to find connections with the exclusively soft-bodied Ediacaran fossils and modern animals, no certain relationships have been established. It is only in the Cambrian period, just over 540 million years ago that the first fossils that are definitely related to modern groups suddenly appear.

In fact, nearly all the modern phyla appear more or less at once as fossils of this age, with such abruptness that the event is known as the Cambrian Explosion. Arthropods, molluscs, echinoderms, annelids, brachiopods and chordates (our own group) all appear in the fossil record at this time.

The specimen shown on the previous page is a fossil penis worm (see object 14) called *Ottoia*. It is one of the more common species from one of the world's most famous fossil sites: the Burgess Shale in western Canada, which dates to the time of the Cambrian Explosion. The fossils found here preserve the soft bodies of these animals in astounding detail. Creatures without any hard parts could be fossilised in this way due to a combination of incredibly fine-grained sediment and the specific chemical and environmental conditions. The odds of that happening are miniscule, but every so often, over the course of geological time, the circumstances align to produce such intricate fossils. We are even luckier that geological processes such as subduction (when one part of the earth's crust moves under another one) and erosion have left these rocks untouched, and that they have come to the surface on dry land during our age.

The Burgess Shale is the best-studied site for the Cambrian Explosion, but a more recently discovered fossil bed, the Maotianshan Shales of Yunnan in China, is slightly older. Across these two localities a great wealth of incredible fossils appear. Among them are obvious relatives of modern species but also others that are rather alien to us. From this diversity it is clear that complex ecosystems had evolved, with different animals occupying different ecological niches, from filter-feeders and scavengers to apparently vicious predators.

It has been hypothesised that the explosion of diversity observed among the Cambrian fossils is related to the evolution of complex vision in animals: the oldest image-forming eyes appear at this time. With eyes to see, active predation can begin, and that would inevitably have spawned an evolutionary game of cat and mouse between predators and prey. The innovation of an adaptation for attack brings about the evolution of a defence – here, perhaps, was the start of the first arms race. Could such a set of circumstances account for the sudden evolution of all of these animal groups?

While there is undeniably some logic in that argument, and it could well be true, it is not certain that the abrupt appearance of diversity is anything more than a sudden change in fossil preservation. Could the Cambrian Explosion be a false impression resulting from an incomplete fossil record?

Another fossil site in southern China, the Doushantuo Formation, dates from 580 million years ago – 40 million years before the record of Cambrian Explosion. It too has exceptional preservation, to such an extent that individual clusters of cells have been picked out. It has been suggested that these balls of cells are actually the dividing early-stage embryos of bilatarian animals like echinoderms or arthropods. If so, that would put the appearance of modern animals well before the Cambrian. Alternatively, they could just be dividing bacteria.

There is also molecular evidence that modern body plans appeared earlier than the fossil record suggests; possibly hundreds of millions of years earlier. These estimates come from calculating the rate at which various complex molecules change over evolutionary time, and then comparing how different the molecules are in different groups. It is then possible to extrapolate back to estimate when the two groups diverged from their last common ancestor.

Could it be that the earliest animals didn't leave any fossils for significant periods of their evolutionary history? That could be down to the dissolved mineral content in the seas of that period, or because the constitution of the animals didn't lend itself to fossilisation. Or maybe we just haven't found them yet. Or perhaps the molecular calculations are too prone to error over such long periods of time and the Cambrian Explosion really was a burst in diversity around 540 million years ago, as fossils like *Ottoia* suggest.

In any case, it is a significant point in the history of animals as it is the first time that we can say that the groups of organisms we recognise as modern animals had appeared. From here we can start exploring the major steps in our and their evolutionary stories.

21 Jawless Fishes

EDWARDS PLASTER MODELS

Fishes Without Jaws: Bone Idols

As has been mentioned before, 'fish' is not a very good biological term, as taxonomically speaking all vertebrates are fishes (including us) because the first vertebrates were fishes. Nevertheless, most biologists have chosen to put convenience ahead of taxonomic regulation and I think it's fair to say that everyone knows the intended meaning when we say 'fishes'.

First, a note on grammar: when we talk about numerous individuals of the same species, the plural of 'fish' is 'fish'. When we talk about numerous different species, the plural of 'fish' is 'fishes'.

Plaster models of the Devonian fossil fishes *Cephalaspis*, *Pteraspis* and *Drepanaspis* made by Vernon Edwards as teaching aids from the 1920s–50s.

The earliest vertebrates to arise were largely recognisable as fishes, but they lacked jaws (and also bone). Jaws were an adaptation that came as a later step in the story of vertebrates, and as we shall see was a highly significant one in the diversification of the vertebrate form. However, for a long time fishes did just fine without jaws: jawless fishes were very diverse from the Cambrian to the Devonian periods.

Today, only two kinds of jawless fishes have survived, and ironically they appear to be the oldest and least representative forms. Living hagfish and lampreys are believed to represent groups that branched off very close to the base of the vertebrate tree: they have no armour or hard tissues, no paired fins and no jaws.

They are both somewhat wormlike in their form, but the details of their anatomy are different to each other, as are their feeding habits. Lampreys are freshwater external parasites. They have circular mouths arranged like a suction cup with many rings of recurved horny teeth. They use these to hold firmly on to the side of a live fish. Their tongues are also toothed and act like a piston to rasp away at the skin and muscles.

Hagfish are marine scavengers, but have no less remarkable feeding techniques. First, they latch on to a dead animal on the seabed with rows of horny teeth. By folding these teeth in and out they may be able to rasp off a chunk of meat. However, if that doesn't work they have another means of removing flesh which makes great use of the fact that they are extremely

flexible because they lack bone. They tie their tail in a knot and then pass the knot down their body until it is at the head end, pushed against the animal they are eating. This gives them enough leverage to pull their head backwards through the knot and tear off a mouthful of food. The fact that they are coated in slime (one species is called the slime hag) makes this technique even more effective. They also perform the slime-and-knot trick if grasped by a predator.

Many of the key moments in early fish evolution took place in the Ordovician and Silurian periods, 485–419 million years ago, during which time all the major groups appeared. However the Devonian period that followed is particularly famous for an incredible set of armour-plated fishes which made the most of one of their ancestors' critical developments: bones in the skin.

The conodonts of the late Cambrian period are the earliest vertebrates with any hard tissues. They had scales containing calcium phosphate: the mineral that is found in bone. As such, the earliest uses for bone were as protective armour plates in the skin, and analysis of these plates shows that they had more in common with our teeth than our bones – they even had a pulp cavity like teeth do. Indeed shark scales are very similar structures (they're even shaped rather like teeth) and are known as denticles, or 'little teeth'.

Inside their bodies, the earliest vertebrates had un-mineralised internal skeletons (much like lampreys and hagfish still do), possibly formed of cartilage. In later stages of fish evolution calcium phosphate crystals mineralised the cartilage to form bone. This was clearly a huge development in the evolution of vertebrates, as was the acquisition of paired fins (those that appear on both sides of the body – the pelvic and pectoral fins): the appendages that would later become vertebrate limbs.

The earliest armoured fishes didn't have paired fins. However it is clear from the streamlined torpedo shape of heterostracan fish like *Pteraspis* that they were fast swimmers. Their well-developed muscular tail would have propelled them forward, but their swimming technique must have been somewhat awkward and unstable without being steadied by paired fins.

This giant leap in our evolutionary story was seen in groups that followed. Fishes like the osteostracan *Cephalaspis* had well developed pectoral fins (equivalent to our forelimbs) with complex musculature. It's hard to understate the importance of that event in the history of the vertebrates.

Groups like this had now acquired both bone and paired fins, but they remained jawless. Bony plates around the mouths of many of these tank-like fishes provided a similar function to teeth and jaws. It seems likely that these rods and blocks of oral armour were used to grab prey as they swam through the water.

22 Cookie-Cutter Shark

JAWS

Cartilaginous Fishes: Evolving Features To Get Our Teeth Into

Although bony plates around the mouths of the jawless fishes afforded them some abilities in processing food, it was not until the evolution of true jaws that a new world of feeding opportunities was opened up to the vertebrates.

The development of horizontally opening jaws marks one of the key defining characteristics of the next group to appear in our family tree: the gnathostomata (which means 'jawed mouth'). Although colloquially this group is referred to as the 'jawed fishes', as a counterpoint to the jawless ones that came before, this group actually incorporates all vertebrates on land (including us) as well as the sharks, rays and their relatives, and the ray-finned and lobed-finned fishes. The vast majority of vertebrates are gnathostomes.

The first major group to make use of jaws were the placoderms – a group of armoured fishes that appeared in the Silurian period and were diverse until their extinction at the end of the Devonian period. In many ways they were similar to the tank-like jawless fishes that lived around the same time – with parts of their bodies coated in thick plates of bone. Aside from jaws, these early gnathostomes had made another major evolutionary leap forward: they had pelvic fins as well as pectoral ones.

With these two sets of paired fins – one near the front and one near the back, swimming vertebrates could finely control their movements through the water, a feat that was no doubt useful for both hunting and avoiding predators at speed.

As handy as jaws clearly are for feeding, it was not actually their original purpose. Gnathostome jaws are modified gill arches – the structures that support the gills in earlier fish groups. As fishes became faster and more active they had an increased need for oxygen. The gill arches became mobile which allowed them to actively pump more oxygen-carrying water over them, and eventually this led to the front-most arches being repurposed to handle food. Therefore, the first gnathostomes had jaws but no teeth; they came later, after groups like placoderms began shearing and crushing food with their modified gill arches.

The diversity of the next group of vertebrates to appear in the fossil record – the cartilaginous fishes – owes a lot to the range of adaptations seen in their teeth. The approximately 700 living members of this group (including rays, skates, sharks and chimeras) and their extinct relatives have evolved a wide suite of different teeth which allowed vertebrates, for the first time, to really experiment with food handling.

Even just among modern sharks there is a variety of tooth shapes, depending on their diet. Broad, serrated teeth can cut up larger prey like a steak knife (up to whale-sized animals); sharp, pointed teeth allow sharks to grip fast-moving animals like squid and smaller fishes; and species with sets of molar-like crushing teeth can break into shelled molluscs and crustaceans. The largest species of all – whale sharks and basking sharks – have actually foregone the need for teeth and feed on plankton and small fishes that they sieve through their gills in huge quantities. In other groups the presence of teeth allowed vertebrates to delve into herbivory for the first time.

The species depicted here – the cookie-cutter shark – has huge teeth for its size, and it uses them in a remarkable way. By sucking onto larger

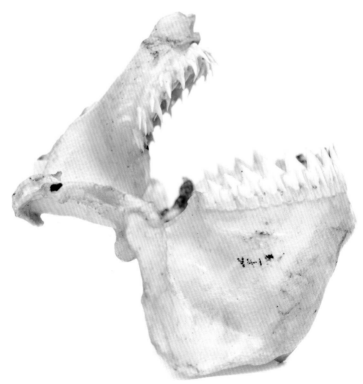

fishes and marine mammals with its fleshy lips, it bites on and twists its body until an almost perfect circle of flesh is removed. It's more of a parasite than a predator as the wounds are not fatal. Cookie-cutters' bellies are covered in bioluminescent organs that camouflage the shark from below, matching the light coming from the sky above in an adaptation called counter-illumination. An area around their chins is left dark, and it is hypothesised that this acts as a lure to large predators who mistake their chin-patch for a smaller fish. When they approach to attack, they find themselves falling victim to the cookie-cutters' circular jaws. While many deep-sea animals use bioluminescent lures to attract prey, this is believed to be the only example of a predator attracting prey by mimicking the *absence* of light.

Modern sharks are famous for their conveyor-belt-like replacement of teeth. As they become used and worn they simply fall out and another one growing behind rotates around the jaw to come into use. In this way sharks can go through tens of thousands of teeth in their lifetime (which can be very long – at nearly 400 years old the Greenland shark is the longest-lived vertebrate known), and explains why shark teeth are such abundant fossils. Incidentally the cookie-cutter shark is the only species known to replace a whole row at a time, presumably as they use them as a single unit. In other species teeth just fall out individually. Running out of usable teeth is a significant disadvantage for many animals (including humans), and one might wonder why the sharks' replacement method isn't seen in more groups of animals.

The cartilaginous fishes form a group called Chondrichthyes which first appeared in the early Silurian period. Instead of bones, their skeleton is formed of cartilage (the material from which our noses and ears are made) which is reinforced by little prisms of calcium phosphate. It is far lighter and more flexible than bone.

They may be light, but as they don't have swim bladders like bony fishes, they have to maintain their buoyancy. They have oil-rich livers which help in this regard, but their rigid pectoral fins act as hydrofoils – as they move forward the forces from the water moving over the fins create dynamic lift. It means they need to swim to avoid sinking.

The prey-detecting adaptations of sharks and rays are astounding. Like other aquatic gnathostomes they have a lateral line system that detects changes in water pressure to provide information about the movements of prey. However in addition they have the most elaborate system known for sensing the tiny electric impulses given off by muscles in their prey. They also have a directional sense of smell: by comparing the difference in the time that a scent reaches their right and left nostrils, they can calculate which way a smell is coming from in the same way that our ears can pinpoint sounds.

23 Common Carp

The Ray-Finned Fishes: A Bony Bonanza

Not long after the appearance of cartilaginous fishes in the Silurian period, the bony fishes came along. In addition to the cartilage encased in a mineral layer seen in earlier groups, bony fishes – called Osteichthyes – have bone that is mineralised with calcium phosphate throughout, and grows from cells embedded within it. This is the bone of your own skeleton.

Soon after the bony fish lineage appeared it split to follow two evolutionary trajectories: the lobe-finned fishes (sarcopterygians), which have muscular fins with a single bone at the top (they are our ancestors); and the ray-finned fishes (actinopterygians), which have fins supported by a series of fine flexible rods. It is the ray-fins that have gone on to be the dominant vertebrates in the seas, lakes and rivers: there are around 30,000 living

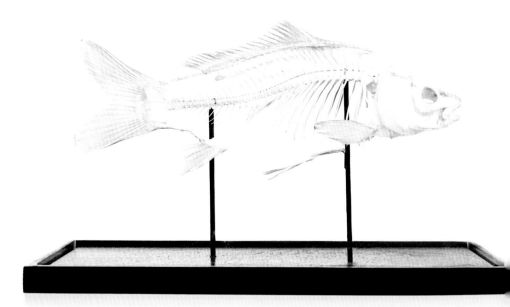

species, making them by far the most diverse vertebrate group. Salmon, cod, guppies, goldfish, tuna, sturgeon, plaice, seahorses, perch, bass, and clown fish are all ray-finned fishes. Think of a living 'fish' that isn't a shark, ray, chimera, lungfish or coelacanth and it will be a ray-fin.

The sea is a big place. It's a lot more three-dimensional than other environments (at its deepest – in the Pacific's Mariana Trench – it's around 11km deep); there is a huge range of habitats as one ventures through the depths, across the latitudes, and along the different kinds of shore. Other aquatic environments on land (such as lakes and rivers) also differ depending on the climate and ecosystem of the world around them. This means that fishes have a lot of different niches that they can adapt to, which helps explain some of their diversity. On top of that, ray-finned fishes have made some modifications to the way they move and feed which has allowed evolution to get to work on producing many ways to be a fish.

A trend in the evolution of ray-finned fishes is that the paired fins (the pelvic and pectoral fins that stick out roughly sideways) tend to become smaller and more manoeuvrable through time. Looking at early fossil members of the group, like *Cheirolepis* from the Mid Devonian period in Scotland, its fin rays are covered in dermal bone and are very immobile (pictured overleaf). Looking at a more modern species like trout, the fins are small and lack a bony covering. This reflects the fact that early gnathostomes (jawed vertebrates) had to generate lift using their fins as hydrofoils to stop them from sinking, like an aeroplane wing, but in water. While other groups developed different means of counteracting this – like the oily liver of sharks – many bony fishes developed a lung, from which the ray-finned fishes have evolved swim bladders.

These structures allow the fishes to remain motionless in the water without sinking, and so the paired fins were no longer required as hydrofoils and could be reduced and modified. These smaller fins enable many ray-finned fishes to have a great deal of agility in the water. Combined with speed this can make them very efficient predators.

By the late Permian period (around 250 million years ago) more advanced groups of ray-finned fishes had freed up much of the bone in their skulls. This was probably the most significant event in the evolution of modern fish feeding mechanisms since the development of the jaw. Detaching the bony elements of the jaw from the cheek and developing the associated muscles allowed the jaw to be swung forward when the mouth opened (picture a goldfish pumping its mouth). As the mouth is pushed into a rounded gape, a region of negative pressure is developed inside it and this causes suction, pulling food into the mouth.

Further developments still are seen in the teleost lineage of ray-finned fishes. This is the group that accounts for the vast majority of fish diversity. More or less every species that isn't a sturgeon, bowfin, bichir, paddlefish

Early ray-finned fishes, like *Cheirolepis* from around 400 million years ago, had rigid coverings on their fins.

or gar are teleosts. Some teleosts separated the muscles that close their jaws into separate bundles, which all contract at slightly different times, increasing the power of the bite.

The diversity of teleosts – covering the shallow seas, ocean depths, rainforest rivers and mountain lakes and pretty much every other watery habit across the globe – is arguably a consequence of the evolutionary potential resulting from the way they have modified their skulls and fins. The important steps in their history include the evolution of jaws by their ancestors; the increased mobility of their fins; the arrival of the swim bladder, which enabled more precise movements; the freeing up of bones in the jaw; and the development of mechanisms to increase the power and efficiency of the bite. The teleosts are a huge and global group, their success is no doubt related to their specialised modes of moving and feeding.

24 Coelacanth

FOSSIL NODULE

Lobe-Finned Fishes: These Fins Were Made For Walking

You are a lobe-finned fish.

We know this because you have a single bone that connects your limbs to your body: the humerus in your arms and the femur in your legs. This feature is one of the key characteristics of the lobe-finned fishes and first appeared in the late Silurian period, around 418 million years ago, when bony fishes split into two evolutionary pathways: the ray-finned fishes (the actinopterygians – see the last chapter), and the lobe-finned fishes (the sarcopterygians).

Today, lobe-finned fishes comprise the lungfishes, the coelacanths and all the four-legged vertebrates on land – the tetrapods – which are the mammals, birds, reptiles and amphibians (and their relatives that returned to the sea, including whales, sea cows and turtles). So really, only a few species of lobed-finned fishes are still actually 'fishy' fishes, the rest are the tetrapods.

The next chapter will focus on how some lobe-finned fishes were able to colonise the land, however it's important to note that if the single bone of the 'lobe-fin' had never appeared, the first tetrapods would never have been able to bear their weight on the fins/limbs on land, so we would never have evolved.

Of the lungfishes and coelacanths, it is the lungfishes that are more closely related to tetrapods. Today, only six lungfish species survive in the fresh but muddy waters of South America, Africa and Australia. This suggests that they are a relic of the time when all the great southern continents were united in the super continent Gondwana, but actually this is just a coincidence, as lungfish fossils are found all over the world.

Both lungs and swim bladders have evolved as out-pockets of the gut, but their relationships to each other are complicated as at some points lungs have been modified into swim bladders, and occasionally later turned back into lungs. No vertebrates have both a swim bladder and lungs, as they are variations of the same feature. In any case, like the tetrapods on land, lung-fishes have lungs. When there is too little oxygen dissolved in the water for their gills to extract, they use their lungs to breathe air.

The ability to reduce the fishy link to water has gone even further in the South American and African lungfishes: they can inhabit water bodies that periodically completely dry up by encasing themselves in a cocoon of mucus (somewhat reminiscent of sci-fi monsters). They will burrow into the mud and eventually, once it has been completely baked dry just sit the season out in a kind of suspended animation, connected only to the outside world by an air tube. When the wet season brings the rains to refill their pools, the water dissolves their mucus casing and they awake from their torpor.

Fossil coelacanth *Coelacanthus sp.* from the Triassic of Greenland. This fossil genus was discovered before the living coelacanths.

Modern coelacanths showing how the fleshy anal and second dorsal fins resemble the paired fins.

Modern coelacanths are large, fleshy lobe-finned fishes from the depths of the Indian Ocean. They have modified their lungs to act as buoyancy devices that function in the high-pressure environments of the deep sea: rather than being filled with air (which firstly they would never have a chance to fill, and secondly would collapse in their underwater world) they are filled with a waxy fat.

Coelacanths have an extensive fossil record and the extinct species also did something different: their gas-filled lungs were lined with very thin layers of corrugated bone – with the corrugations running in different directions which made it incredibly strong. This created a swim bladder of fixed volume – a feature only seen elsewhere in cephalopod molluscs like ammonites. With a solid gasbag like this, however, these fossil coelacanths would have been restricted to shallower seas.

On occasion, scientific research can come up with something so amazing that it stays with you forever. For me, one such moment came at university when reading an article by the Swedish palaeontologist Per Ahlberg in the journal *Nature*. It was less than a page long. He noted that the unpaired fins along the back and belly of most lobe-finned fishes are supported by a row of parallel rods, as they are in most other fish groups. However two of the unpaired fins in coelacanths have a skeleton, musculature and nerve connections almost identical to what is seen in their paired fins. Not only that, but these fins move in sequence with the paired fins. This would suggest that there was a switch in the locations that the genes controlling for paired fins are expressed in the coelacanth embryo: a mutation occurred that built extra pelvic fins instead of normal unpaired fins. What's remarkable about this is that it is an evolutionary change that would have happened suddenly – between generations. A single male and female coelacanth would have produced a baby or babies that had features not seen in either of them. Even more remarkable is that this single one-off event then changed the course of coelacanth evolution forever: it is the only condition found in subsequent coelacanth species.

25 *Ichthyostega*

STARLUX PLASTIC MODEL

Tetrapods And The Evolution Of Vertebrate Life On Land

How did we get here? So far we have explored some of the things that happened on earth that made it possible for organisms like us to evolve. I'm certainly not suggesting that everything that preceded this point in history happened *in order that* we could exist – evolution is not progressive – but it is a fact that there are many developments in our evolutionary past without which we would not be here.

Although by the time our ancestors crawled out of the sea the arthropods had already found myriad ways to make the most of life on land (indeed perhaps the feast of insects and millipedes is what attracted our fishy forebears up the beach in the first place), it is no exaggeration to say that the colonisation of land by vertebrates is one of the most significant events in global history.

Vertebrate life on land evolved from fishy things turning into amphibiany things. What I'd like to consider is what evolutionary changes fishes went through which made it possible for them to survive on land, and become tetrapods – the land-based group which contains amphibians, reptiles, mammals and birds.

One of the most charming collections in the Grant Museum of Zoology is 100 or so models of extinct animals ranging from fishes to dinosaurs. They were purchased in the 1970s and 1980s in order to help students understand what the whole animal looked like when they were studying individual fossils (today we have Google). Incredibly, the curators at the time saw fit to formally accession them into the collection: they have official catalogue numbers and everything. One of them is of the tetrapod *Ichthyostega*. *Ichthyostega* is a very fishy tetrapod.

In order to consider the characteristics necessary for life on land, it's important to examine what characteristics the animals that gave rise to tetrapods had, since evolution can only work on things that are already there. On the branch that led to tetrapods are the 'osteolepiforms', like *Eusthenopteron* (pictured overleaf). This fishy animal had a pectoral fin with a simple humerus articulating with a radius, an ulna and the 'shoulder' (if fishes have shoulders); no real digits; but crucially – different from what came before – is the presence of an interclavicle bone, which gives stability to the limbs' frame by connecting them firmly to the ribs. These are all precursors to moving on land. As in other fishes, its head was solidly attached to the shoulder girdle.

For a fish like *Eusthenopteron* to evolve into something that could walk on land, what changes were required of the skeleton? The most obvious character is that legs need to develop from fins. Locomotion on land is vastly different from that in water in three main ways.

First, it must be entirely limb-driven: while swishing your body sideways like a fish can increase the stride length on land (think of lizards and crocodiles with their sprawling gaits), the force of the step comes from the legs. Compare this to the tail-driven swimming of fishes.

Second, moving on land is far more energetically demanding: the body is not being supported by water, so the role of the skeleton in holding the weight must be increased.

Third, gills are of no use on land, and respiration must occur either by absorbing oxygen through the skin, or pumping it into the lungs using the mouth, throat or ribs.

If you squint you can see solutions to these needs on the Starlux model of *Ichthyostega*, one of the very first tetrapods from the early Devonian period. This animal had four legs. And digits – fish fingers! On the forelimb is a small point of bone that sticks out in the elbow, which is called the olecranon process. This is where the muscles used to straighten the limb are attached to the joint, and it's the thing that stops you from bending your elbow in the wrong direction. It is critical for bearing body weight on all fours. All in all, *Ichthyostega* shows a good mix of fish- and tetrapod-like features.

A massive advance from the situation in *Eusthenopteron* is that the bones that support the hind fins/limbs are *attached* to the spine and thus

Starlux model of *Eusthenopteron*, a lobe-finned fish with features that would allow its relatives to eventually colonise the land.

provide some solid support – just like the shoulder girdle connects the pectoral fins/forelimbs to the spine. In earlier fishes the pelvic bones just floated in muscle, but in *Ichthyostega* they connect to the vertebrae and grow down and around to meet under the belly, forming an all-round pelvic basket. In this way the hind limbs – which are larger than in their fishy ancestors' – can support the animal's weight under gravity on land.

Ichthyostega's ribs are unusually massive. They are broad and overlap considerably to form a near-solid wall. This supports against the effects of gravity, and provides sites for muscle attachment for flexing muscles.

The difference that I think most people miss between fishes and tetrapods is that fishes don't have necks. Why is that important? In *Ichthyostega* the shoulder girdle is no longer fused to the head (you can tell this because the Starlux model has a wrinkly neck). This stops jarring vibrations having such an impact on the brain and skull, and frees up any ties between locomotion and feeding. Try wearing a neck brace whilst walking and drinking a cup of tea to see why that's important.

Ichthyostega has also developed new joints between adjoining vertebrae, called zygopophyses, which means the weight held by one vertebra is held up by the next one, spreading the weight all down the spine.

It is through these adaptations that fishes were able to walk onto the land, turn into early tetrapods and then go through a number of steps until we finally get platypuses (and also humans). Thanks *Ichthyostega*. I mean it.

26 *Wellesaurus*: A Triassic Amphibian-Relative

PLASTIC CAST OF SKULL

The First Tetrapods

With the evolution of the features enabling fishy animals (technically known as fishapods) to venture up onto the shore or riverbank came the arrival of tetrapods. These first tetrapods, like many of their descendants (those who have not since lost these traits), had digits, a pelvis solidly attached to the spine, supportive joints between neighbouring vertebrae and a neck freeing their heads from their shoulders. These adaptations

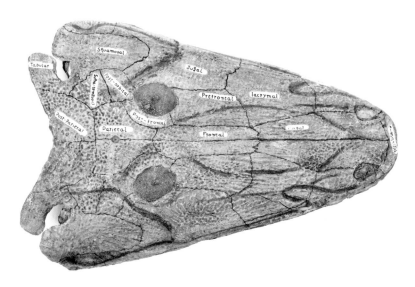

allowed animals like *Ichthyostega* to become so successful at using the land that their descendants have joined the arthropods in taking it over.

The earliest tetrapods would have only been able to venture onto land for relatively brief excursions. Indeed, it's clear from fossils that *Ichthyostega* retained the gills of its fishy ancestors. The animals that immediately followed were amphibious and maintained a mandatory link to fresh water – the last obstacle to overcome when colonising the land was how to lay eggs that didn't dry out when exposed to air. The early tetrapods, like the vast majority of living amphibians, had to return to the water to breed. An evolutionary solution to this challenge was found by the amniotes – a group that includes mammals, birds and the animals we traditionally called reptiles – which protected their eggs from desiccation by surrounding them in specialised membranes (see the next chapter).

Historically the word 'amphibian' had been used to describe the early tetrapods broadly, as well as the three groups that we find alive today: the frogs, salamanders and caecilians (enigmatic tropical burrowing animals that resemble worms). However, some of those diverse fossil groups were more closely related to us amniotes, and so this use of the word becomes troublesome. Therefore to avoid confusion with the general term 'amphibian', zoologists call this modern group the lissamphibians, from the Greek for 'smooth amphibian', referring to their skin (but actually caecilians do have scales so that doesn't really help). Fossils described in 2017 suggest that this group originated before 315 million years ago.

Lissamphibians comprise around 7600 species today, all but 900 of which are frogs and toads. Nearly all of the animals that have been covered so far in this book have a larval stage – the animal that hatches or is first born is anatomically quite different from the adult, and they go through a metamorphosis to transition between the two stages. This life history isn't seen in the amniote group that evolved from amphibious ancestors – the mammals, reptiles and birds – but the lissamphibians are famous for it. Every schoolchild knows that frogs start off life as tadpoles (but for the sake of accuracy it's worth saying that some frogs skip the tadpole stage and emerge from the egg as froglets).

These larvae retain gills and are on the whole aquatic, while nearly all of the adult forms can take oxygen directly out of the air. It is therefore interesting that lissamphibians have either shrunk or completely lost their ribs, so they can't expand their chests to fill their lungs like we can. Instead, a lot of their respiration takes place directly through their skin. This is why amphibian skin needs to stay moist: so that gases can dissolve across it. It is this requirement that has kept amphibians out of the driest places on earth: the poles. Those that live in deserts use behaviours like burrowing and cocooning; they only emerge in periods of rainfall (I was once in Australia's Tanami desert during a rare rainstorm and the iron-red sand dunes erupted with frogs blistering out of the ground). They also pump air into their lungs using their throats and mouths to breathe, as any cartoon frog will demonstrate.

When people think of modern amphibians most animals that come to mind are rather small – creatures that would easily fit in one's palm. Frogs that surpass 'normal' frog sizes are often met with surprise, even though they are barely larger than a grapefruit (the modern record is the goliath frog at over 3kg, but in the Cretaceous period, *Beelzebufo* was half as big again, and may have eaten small dinosaurs). Nevertheless the largest modern amphibian can reach an impressive 1.8m long – the Chinese giant salamander.

'Giants' like this were certainly not unusual in the fossil record. From the Carboniferous period to the early Cretaceous period. amphibians had very different dimensions from modern species. In their day there were species that closely resembled crocodiles, with large flat triangular heads packed full of pointed teeth (some even had tusks) and long bodies with a deep, narrow tail adapted for swimming. The largest species, including *Prionosuchus*, surpassed 5m long.

One of the most significant groups was the temnospondyls, which is the strongest contender for the honour of being most closely related to (or even the direct ancestors of) modern lissamphibians. Perhaps the best-known species was *Eryops* from the Permian of America, around 300 million years ago. To my eye, it looked like the cross between a crocodile, with a long flat tail and croc-like legs; a pig, with a barrel-shaped body (though being an amphibian, it had short ribs); and a hippo, with a large, deep skull (albeit full of long needle-like teeth). It weighed around 90kg – the size of a very large boar.

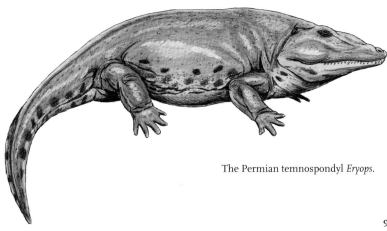

The Permian temnospondyl *Eryops*.

27 *Diadectes*: A Close Amniote Relative

PLASTER CAST OF SKULL

The Amniotes: Severing Ties With Water

The soft, jelly-like eggs of amphibians have a significant weakness: they dry out easily. This means that they cannot be laid on land, and so amphibians can never venture far from fresh water if they intend to breed. Amphibians' thin, breathable skin also prevents them living in salty environments like the sea, as they cannot tolerate the salt imbalance between the inside and outside of their bodies (this problem has been solved by one modern species, the crab-eating frog, which stockpiles urea to make up for this difference in salt concentrations).

The jelly egg capsule of amphibians also puts a size constraint on the animal that can emerge from it. That nearly all amphibians have larvae is an outcome of the fact that amphibians cannot grow big enough in their eggs to complete their development into adult form. Lacking a structural membrane, jelly eggs cannot exceed a certain size without collapsing under their own weight. If the egg were to be increased in size, a large amount of jelly would also prevent enough oxygen from reaching the developing embryo. This is because the relative amount of surface area through which to absorb oxygen decreases with increasing volume.

A group of animals arose with solutions to these problems: principally in the form of the amniote egg. By coating their eggs in breathable but protective membranes, these animals could cut their reproductive ties with the water and venture permanently onto dry land. These membranes can be seen, for example, just below the hard shell of a hen's egg, surrounding the albumen of the egg white. The amnion or amniotic sac of us mammals is the same structure (this is what bursts when one's waters break), it's just that we no longer lay eggs. These fibrous membranes provided the structural support to allow the egg to grow large enough for adult forms to develop directly in the egg. The transfer of water from the jelly-like albumen to swell the yolk may also push the embryo close enough to the membrane for gas exchange. Released from the constraints of the jelly capsule, the egg could increase in size.

The amniote animals that first evolved are regularly referred to as 'reptiles', but like so many terms in common use, this word causes problems, as technically it would have to include mammals (which evolved from early fossil members of this group) and birds (which later evolved from dinosaurs). The zoological term for mammals, birds, lizards, snakes, crocodiles, non-avian dinosaurs, turtles and all their fossil relatives together is the amniotes.

The problem with defining a taxonomic group with a feature made of soft tissue – like the thin amniotic membrane of amniotes – is that it doesn't fossilise. This means that it's very difficult to work out when in the fossil record the group arose, and therefore what specifically they evolved from. This may seem relatively unimportant, but for whatever reasons, palaeontologists and taxonomists have a penchant for being able to say 'this fossil is the first ever vertebrate/reptile/dinosaur/mammal/bird etc.' They have always struggled to do this with amniotes, which is particularly frustrating for them as it's such a significant evolutionary step.

To attempt to get around this, palaeontologists have sought to find bony features that unite all amniotes which they would be able to find in the fossil record. For example, all living amniotes have a special arrangement of ankle bones – including a characteristic astragalus bone (the main bone joining your shin to your foot) – key adaptations to a life walking on land.

The problem is that it is uncertain that the change in ankle anatomy happened at the same time as the change in reproductive development. This makes it effectively impossible to define the first members of the amniote group from the fossil record.

Diadectes, pictured on p. 94, was once considered to be an early amniote, but as an example of these taxonomic troubles, it is now regarded as a close relative of amniotes, on a neighbouring branch of the tree. It had an interesting mix of amniote and non-amniote characteristics, such as strong reptile-like ribs, which are essential for sustained weight-bearing on land, with a relatively amphibian-like skull. However, looking at its teeth, it was clearly herbivorous – a lifestyle not commonly associated with amphibians. This raises interesting questions about how it digested plant-matter. This would presumably have only been possible with the help of symbiotic gut bacteria.

Animals are not born with a pre-existing ecosystem of bacteria in their gut, so they need to acquire them very early in life, typically either from their mother's birth canal or from eating poo. It's hard to picture how that would have happened if baby *Diadectes* started life in the water. Without the early phase of life being in water, amniotes do not have a larval stage like their immediate ancestors. Fossil larvae of *Diadectes* have never been found, raising the possibility that it had some kind of transitional reproductive tactic, somewhere between amniotes and what is still seen in modern amphibians, or even that it live-birthed its young.

Whatever the first amniote was, the evolution of land-based reproduction was an extraordinary development in the history of the earth, at least from a human point of view, as without it there would be no mammals, no birds, no crocodiles, no lizards, and no dinosaurs.

The Permian tetrapod *Diadectes*, from around 280 million years ago.

28 *Lystrosaurus*

FOSSIL SKULL AND TORSO

Making Mammals

Soon after the arrival of amniotes the group began to diversify, presumably as a result of their new-found freedom from water. Among them was the group that would give rise to mammals. It is called the synapsids, appearing around 300 million years ago in the Carboniferous. Mammals sprung forth from this group some 90 million years later. Today the mammals are the only synapsids to have survived, but over the course of their evolution there were a great number of non-mammalian synapsids, including *Lystrosaurus*, pictured here.

The major amniote groups are defined by the number of holes at the back of the skull, called temporal fenestrae ('windows on the skull's temple'). The synapsids have only one on each side, but it is quite hard to spot in modern mammals as the brain has expanded to such an extent that it has more or less filled in this diagnostic hole. Our cheekbones form the lower margins of our temporal fenestrae.

The skull is on the left of this object, with a large round orbit, and two short tusks poking out of a their deep, curved roots.

The other major group of modern amniotes are the sauropsids, which have two temporal fenestrae. These are the lizards, snakes, crocodiles, dinosaurs (including birds), pterosaurs and extinct marine reptiles. It is believed that temporal fenestrae arose as they provided *edges* of bone for muscle to attach to, which are much stronger anchoring sites than flat surfaces. The hole may also allow contracting muscles space to bulge. Turtles and tortoises have no temporal fenestrae (the 'anapsid condition') and were once considered to be a third amniote group, but they are now widely held to have evolved from sauropsids, losing their fenestrae in the process.

The most famous non-mammalian synapsids are the pelycosaurs. If you want to watch a palaeontologist lose their temper, give them a mixed box of 'dinosaur' toys. Invariably these include the pelycosaur *Dimetrodon*, the great sail-backed beast with a large, toothy head. For some reason, many toy-manufacturers insist upon including *Dimetrodon* with the dinosaurs, despite the fact that it pre-dates dinosaurs by 65 million years, and isn't at all closely related.

Mammals are defined by producing milk for their young and being covered in hair. Like the amniotic membranes in the last chapter, these are soft-tissue features that do not readily fossilise, which can cause palaeontologists problems. There are some bony features which differentiate mammals from other amniotes, but these were acquired gradually over the course of synapsid evolution.

We are taught in school that the smallest bones in the human body are the bones of the middle ear: the malleus, incus and stapes. They translate sound from the outside world, via the eardrum, into auditory information for our brains. Other than the stapes, we do not share these structures with reptiles; not exactly, anyway. The malleus, incus, and the ring of bone that holds the eardrum evolved from bones of the reptilian lower jaw.

Reptiles like crocodiles and lizards have seven bones on each side of their lower jaws, mammals have just one: the dentary – the one that corresponds to the front of reptilian jaws and holds the teeth. Over the course of synapsid evolution, the fossil record – including in fossils like the therapsid *Lystrosaurus* – shows the shrinking of the bones at the back of the jaw, including the angular (forming the mammalian tympanic ring), articular (malleus) and quadrate (incus). Eventually they are incorporated into the middle ear. Given that reptiles use some of these bones to form the joint between the lower jaw and the skull, this meant that mammals and their closest relatives had to form a new kind of jaw joint for the dentary. The reptilian jaw joint is actually now inside our ears.

A more visible difference between reptiles and mammals is the way they walk: the former have a sprawling gait, with bent elbows and side-to-side movement, compared to the upright stance of most mammals, which bend their spine up and down as they move.

The sail-backed pelycosaur *Dimetrodon*, living between 295 and 270 million years ago.

Side-to-side bending cannot aid pumping of air into the lungs: as they bend to the right, air in the right lung is squeezed out into the left, and *vice versa*. This means that breathing is not possible at the same time as walking or running in animals with a sprawling gait (unless they come up with some other method of getting air in – monitor lizards pump their throats). This means that when tetrapods first walked onto the land it would have entailed a significant decrease in stamina for sustained movement. It would seem that tetrapods primitively have a high capacity for anaerobic metabolism (activity without sufficient oxygen – the kind that gives us a stitch): lizards, for example, intersperse running with brief pauses for pumping their lungs.

In running mammals lung-pumping is linked to the stride cycle: rather than bending from side-to-side they flex their spine in a head-to-toe motion – think of a cheetah bounding – this forces air in and out of both lungs at once (in bats and birds, breathing is similarly tied to wing beats).

Although the maximum speeds of lizards and mammals do not differ, it is believed that the increased capacity for sustained running was the principal factor in the evolution of warm-bloodedness: to be warm-blooded you need such a high metabolism that you cannot stop breathing for any length of time. Throughout their evolution, synapsids – on the route to the evolution of mammals – show the necessary stepwise changes to bring about this shift in modes of locomotion and breathing.

29 Horse Relatives

How A Changing World Produced Modern Mammals

In the period that immediately followed the extinction of non-avian dinosaurs (birds are dinosaurs which survived the mass extinction), the world underwent a series of global environmental changes in which lineages of archaic mammals disappeared and were replaced by most of the modern mammalian families.

The name of that period of the earth's history is the Palaeogene – the 40-million-year spell that spans the Paleocene, Eocene and Oligocene epochs. At the beginning of the Palaeogene, 66 million years ago, the world looked very different to how it does today: a body of water – the Turgai Straits – separated Europe and Asia, which were both distanced from Africa by the Tethys Sea. Greenland linked North America to Europe, and there was dry land joining North America to Siberia. South America – still separate from North America – was joined to Antarctica, which in turn was periodically in contact with Australia.

At the start of the Palaeogene the world was warmer: thick sub-tropical forest spread to within polar confines. The early Paleocene mammals were mainly small and arboreal (tree-dwelling), because the forests were so dense. At this time the low seasonality drove plants to be extremely fibrous, and no mammals were truly herbivorous.

Across the northern continents, the dominant carnivores were called 'creodonts' – a group of medium-sized mammals with general predatory adaptations; and the key hoofed mammals were called 'condylarths'. (It is almost certain that at least one or perhaps both of these does not form a true taxonomic group; they are made up of a mix of animals that taxonomists have historically found difficult to place.) In South America, marsupials and relict monotremes – the group which today includes platypuses and echidnas – wandered the forest with the ancient relatives of sloths and armadillos, and a group of stocky hoofed mammals called meridiungulates.

At the end of the Paleocene epoch, around 56 million years ago, warming began and extended into the next geological epoch, the Eocene, and tropical forest expanded further into higher latitudes. There were still many arboreal mammals and the early primates that originated in the Paleocene flourished. However there is evidence that the understorey was more open at this time, which permitted a diversification of ground-dwelling mammals with the greater availability of edible leaves as food.

This new habitat was ideal for the newly arrived horse relatives – *Hyracotherium* and *Propaleotherium* – which were miniature compared to modern horses. Up in the trees, early lemurs made the best of the fruit, nuts and leaves on offer, as did the newly arising rodents. The diversity of odd-toed hoofed mammals was booming, including horse and rhino relatives. In contrast, the even-toed hoofed mammals (which today include antelopes, deer, camels and pigs) were restricted to very small, predominantly omnivorous forms. The success of the odd-toed hoofed mammals was probably down to their hindgut digestive system, which enabled them to break down fibrous foliage.

The ancestors of modern horses were suited to environments with thicker vegetation. During the evolution of the modern *Equus* lineage, the number of toes decreased while the limb bones increased in length, as adaptations to increasingly open habitats in which they could gallop. *From top to bottom: Hyracotherium, Merychippus, Pliohippus* and *Equus*.

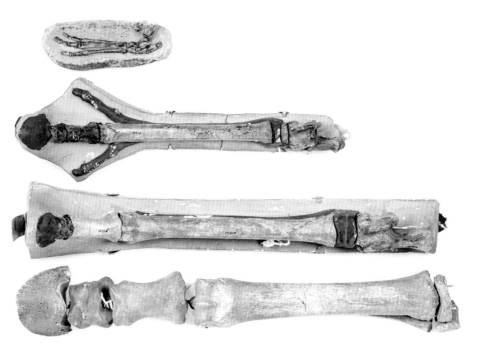

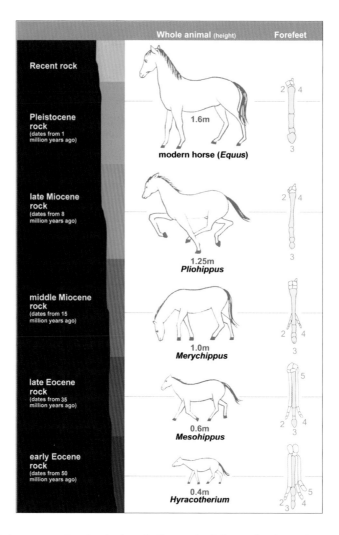

	Whole animal (height)	Forefeet

Recent rock

Pleistocene rock (dates from 1 million years ago)

1.6m
modern horse (*Equus*)

2 4

3

late Miocene rock (dates from 8 million years ago)

1.25m
Pliohippus

2 4

3

middle Miocene rock (dates from 15 million years ago)

1.0m
Merychippus

2 4

3

late Eocene rock (dates from 35 million years ago)

0.6m
Mesohippus

5

2 4

3

early Eocene rock (dates from 50 million years ago)

0.4m
Hyracotherium

5

2 4

3

An artist's reconstruction showing how the lineage that led to modern horses increased in size while reducing the number of toes.

As in the Northern Hemisphere, arboreal and climbing omnivores and insectivores decreased in diversity in the Southern Hemisphere, and the diversity of ground-dwelling omnivores and herbivores increased in their place. In Africa, elephant relatives, hyraxes and sea cows radiated. The world was hot and humid and extremely species-rich. All of this was about to change.

The Eocene grew gradually cooler, and at its close, around 34 million years ago, cold water from the Antarctic spread northwards and the global temperature plummeted. As the ice cap grew the sea level dropped, and

Asia and Europe merged as the Turgai Straits drained. This had profound effects on mammalian wildlife, and most of the archaic forms died out. However, their losses were more than balanced by new types evolving. Among the casualties were the 'condylarths'. More primitive primates were unable to compete with the new arrivals in the cooling world and either deceased in diversity or became extinct.

Herbivore diversity was enriched by the radiation of the even-toed hoofed mammals, comprising three key lineages – the pigs, the camels and the ruminants: the antelope, sheep, giraffes, deer and their relatives which have a specialised fore-gut to ferment vegetation before digestion. Horses, rhinos and their relatives were still around, but the efficient digestive system of the ruminants was too effective for them to compete, and their diversity decreased.

The Oligocene epoch which followed was cooler and drier than the earlier periods of Palaeogene. Broad-leaf deciduous woodland spread across the middle northern latitudes, and there is some evidence of savannahs at this time. With these new open habitats came a new type of herbivore, with long toes and thin legs – adaptions that increased stride length, allowing them to run at speed and travel long distances, probably in herds. The increased seasonality encouraged plants to grow leaves that favoured the ruminant way of life, further replacing the odd-toed hoofed mammals.

Among some hoofed mammals we see a reduction in the number of toes as they adapted to ever more open habitats. Antelopes, deer and their relatives now have only two toes, while modern horses just walk on their one middle finger or toe. The increase in length of each part of the legs allowed for a longer stride – an adaptation to running through more open environments.

The older groups of carnivores were also replaced by cats, bears and other members of their order. At this point many of the animals resemble modern-day counterparts. It seems that the massive change in global climate at the end of the Eocene resulted in reduced temperatures and sea levels. This change was evidently too much to bear for the archaic mammal groups and they were not able to adapt, giving rise or giving way to the animals we see today.

Fig. 5. Dasyure Tafa.

Fig. 6. Dasyure Ursin.

Fig. 4. Dasyure à Pinceau

Fig. 1. Didelphe aux oreilles bicolores

Dasyure Cynocephale Fig. 3.

Fig. 2. Dasyure tacheté

Deseve del. et direx.

Pierron scu.

Histoire Naturelle. Quadrupedes.

Part 3

Natural Histories

HOW DOES EVOLUTION WORK?

'Nothing in biology make sense except in the light of evolution.'* This is probably one of the most quoted biological adages. Some extraordinary things have taken place in the natural world – there are bizarre species, adaptations, behaviours and processes to be seen everywhere we look.

Over the past two centuries or so science has sought explanations for them: how can we explain the existence of an animal that had antlers nearly twice as wide as it was tall? What mechanisms are behind the repeated appearance of similar structures across the animal kingdom? What do we mean when we say two species are related? What is the explanation for certain behaviours that appear to be really costly to an animal's survival? How did species get to where we find them? How and why do species die out? Why do males exist?

In this section of the book I will explore the answers to questions like this by delving into some of evolution's major mechanisms. Stories from ecology (the study of the interactions between organisms and their environment), the science of animal behaviour and taxonomy all intertwine to help us understand the natural world. These topics are some of the central pillars of natural history.

The following objects have been selected to illustrate and explain themes around how evolution works, by exploring for example: natural and sexual selection and convergent evolution; where species live – the science of biogeography; processes underlying animal adaptations; how animals sense the world; some of the genetic systems underlying animal ecology; the different ways that animals interact in intimate symbiotic relationships; and how humans are affecting the world today.

---

* Dobzhansky, T. (1973), 'Nothing in biology makes sense except in the light of evolution', *American Biology Teacher*, vol. 35, pp. 125–129

Left: Several marsupials as they were depicted in 1820 in Anselm Gaëtan Desmarest's *Mammalogie ou description des espèces des Mammifères*. At centre right are a Tasmanian devil (see object 79) and a thylacine (see object 78).

These stories operate on widely different scales, and some of the following chapters will focus specifically on the evolution of a single species, while others incorporate huge ecological systems. For each one, I've chosen an object from the Grant Museum of Zoology to exemplify the theme, but in reality each story could have been told by any number of species, and each species could be used to tell many stories – nature doesn't fit in boxes. Every animal on earth is at the mercy of the same biological process to feed itself, find shelter in its environment and ultimately pass on its genes to the next generation. Discovering how they do it is a joy of natural history.

30 Spiny Stick Insect

PINNED SPECIMEN

How Natural Selection Works

Evolution is change through time. In the context of natural history, discussions about evolution relate to how a lineage has been modified through the course of its history – how new species arise and how the anatomy, physiology and behaviour alter over a series of generations. There are a number of ways in which biological evolution can take place, but the most famous is natural selection.

The idea that species change over time has been in circulation for centuries and had always been controversial. Many felt (and continue to feel) that it challenged a religious view that species were divinely created as they are, and that it was blasphemous (or simply pointless) to search for a

mechanism that could drive change or the development of new species by a means other than God's hand.

Despite this, observers of the natural world had long suggested theories for how changes in species might come about, but none of them were particularly satisfactory. That is, until 1858 when Charles Darwin and Alfred Russel Wallace proposed an idea that has stuck – it neatly explains how, in the simplest of terms, evolution can *work*. The following year Darwin published it in full in a book that is surely near the top of the list for the world's most influential: *On the Origin of Species by Means of Natural Selection*. Despite the clamouring of a powerful religious lobby, natural selection as a principal mechanism for evolutionary change is now considered beyond doubt in the scientific world. It is so beautifully simple that it is essentially self-evident.

Three requirements are essential for natural selection to occur:

1) That there is variation between individuals – not all members of a species are identical.
2) That at least some of that variation is heritable – the variations are passed from one generation to the next. For example, human parents pass on their skin colour to their children.
3) That not all individuals that are born will survive to reproduce. The odds of surviving are linked to the variations between individuals. There is competition; some individuals will live longer because of the characteristics they have inherited from their parents.

All of these conditions are easily demonstrable.

In 1858 the idea of the gene was not yet understood, so people didn't know how characteristics were passed from parent to offspring. Today we know that these variations can be caused by mutations in our genes. Mutations happen all the time – cell division is not a perfect copying mechanism, and the building blocks of our DNA, which determine what our genes do, are occasionally replaced with the 'wrong' building block. Many such mutations will have no implications on an individuals' ability to survive, but some will stop the embryo developing at all, and some might help. Natural selection explains how a new positive mutation can spread through a population by conferring advantage to those that inherit it, and so how a species can change.

In practice, it is hard to observe natural selection or to detect it in the fossil record. We can see change over time, but very rarely at the level of being able to determine one generation from the next. This means that except for species that lived recently enough for us to extract DNA, the proportions of Variety A vs. Variety B in a fossil population at a given time are likely to be impossible to calculate, so any comparison to a later period in time is impossible. In human history, the most famous example is the peppered moth, which was known as a lightly coloured insect adapted to camouflage against lichen on trees. As the industrial revolution polluted the air with soot the lichen died and the trees became darker. A mutation in the moth for a dark dappled form arose, and quickly became established as it allowed moths to camouflage themselves against the newly sooted trees.

Given that this is a very human example, and rather overused, I want to use a theoretical story of how stick insects might have evolved their incredible camouflage adaptations. I'm not suggesting that this is actually what happened, but it seems to me to be a sensible simplified example to explain how natural selection works as a mechanism.

Picture a population of an insect on a tree whose main cause of death is being eaten by predators. The number of eggs an individual lays in its lifetime depends on how long it lives before being eaten. One insect lays an egg with a mutation which results in that offspring growing up to look slightly more like a stick than its parent did. Because it is less readily spotted than its neighbours, it lives slightly longer before being eaten, and so it lays slightly more eggs than the average. Each of this insect's offspring have inherited the mutation, so they also are better camouflaged, and so live longer than individuals without the mutation. They too have more offspring, who also have the mutation, and so they live longer ... and so on.

In this way the mutation to look a bit stick-like spreads throughout the population because more offspring are being produced with the genes for better camouflage. In the long-term, mutations which allow the individual to produce more descendants are going to be favoured by evolution – this is called selection pressure. There is selection pressure for mutations to spread that make the insects even more camouflaged. Over time, ever more elaborate camouflaging mutations spread, and something that looks like a stick insect is the eventual result.

31 Giant Deer

SUB-FOSSIL SKULL

Sexual Selection: Antlers

If natural selection can be summarised as 'survival of the fittest', how is it that some animals have evolved features that seem to be significant hindrances to their survival? Deer antlers, peacock tails and babirusa tusks do not help an animal to stay alive. Darwin asked a similar question in *On the Origin of Species*, and also came up with an answer – sexual selection.

Sexual selection is a sub-set of natural selection, where the driving force is not on the animal to survive, but instead to have the most descendants. It is the mechanism by which species evolve characteristics that help them fight off rivals for access to mates; make them more attractive to the opposite sex; or ensure sexual encounters result in the most favourable outcome (more or fitter babies).

It manifests in a vast number of ways, ranging from one sex having weaponry like horns, tusks and antlers; sheer muscle; the delivery of gifts; males scraping their rivals' sperm out of the females' reproductive tracts (or even the females removing the sperm of previous suitors from their own bodies if a better offer comes along after they've already mated); absurd decoration; building elaborate architectural structures; or undertaking ridiculous dance routines. Generally, if there is a significant difference between the males and females of a species, it's normally safe to assume that sexual selection is the reason.

I'll explore a few of these with future objects, but the world's most visible outcome of sexual selection is in the giant deer, which lived from 400,000 to 7,000 years ago, ranging from Siberia to Ireland (they are also known as Irish elk, but they aren't really elk and aren't exclusively Irish). Standing at around 2m tall, males grew the largest antlers of any deer, reaching 3.6m across – these animals were almost twice as wide as they were tall.

Like in all male deer, these antlers were shed and regrown *each year* to coincide with the mating season. Deer antlers are solid bony outgrowths of the skull's frontal bones (the ones above the eyes), and a large set on a giant deer could weigh 45kg – what the UK government considers 'healthy weight' for a person five feet (152cm) tall. I find the idea that each year an animal could grow – as an extension of its 2kg skull – a mass of bone equivalent to a small adult human frankly astounding. It would have been some of the fastest growing bone in any species – estimated at a rate of 8.7mm per day, including 100g of calcium. A stag would have needed to consume around 40kg of grass, leaves and twigs each day during the growing season in order to feed his antlers. Red deer on Scottish islands are known to eat ground-nesting seabird chicks as a calcium supplement. It's plausible that giant deer could have done similar things.

Deer antlers evolved as a result of competition between males for access to females (a sub-set of sexual selection called 'male-male competition' or 'intra-sexual selection'). Before coming to blows, two males size each other up, displaying their antlers at the best angle to show off their size. At this stage, if one of the males decides that there is a significant disparity between them, and that he has little chance of winning in head-to-head combat (literally), he will back off and leave the larger male to pursue the females. If they look to be equals, however, it comes to blows.

The shape of a deer's antlers have evolved to interlock with another male's as they clash. Animals of the same species very rarely fight to the death (sex is important, but not that important), and so mortal wounds are rare – the arrangement of the antlers' prongs (called tines) are adapted to stop serious injury to the face and body. Indeed many species, including giant deer, have short tines sticking out over their brows to keep their faces out of trouble.

Once locked together, the two battling males push and twist, trying to cause the other to lose balance, or get into a position where one of the outer tines can be used as a dagger, weakening their rival. It's incredible to think of the power that would be needed to wield a 3.6m-wide 45kg weapon as a battering ram, and the anatomy needed to withstand the shock of such blows. This is why the bones of a giant deer's neck have huge vertical projections – to carry enough muscle to keep the head – and its huge appendages – held up.

Depending on the species, winning such a battle could come with serious rewards. By extracting ancient DNA from sub-fossil specimens in museums, as well as by comparing their morphology, giant deer have been found to be most closely related to modern fallow deer. The females of this species tend to gather in groups, so a male who succeeds in fighting off other would-be suitors can end up having access to a large number of mates, and siring a large number of young. In so doing, he is passing on his genes for big antlers and the strength to use them. It is reasonable to assume that giant deer society might have been similarly structured.

For obvious reasons, giant deer are prized objects by museums – a large rack of antlers cannot fail to impress, particularly when they can say that the species once roamed the very region that the museum is now based. It is rare to visit a natural history museum within their former range that does not have one on display. However, they tell a story of inherent bias in museums. Because they don't have antlers, female giant deer are very rarely put on show. As institutions dedicated to communicating science, it is easy to assume that museums would be impartial and 'fair' in their coverage of species in displays. When it comes to animals with very different males and females, however, the less showy of the two invariably gets left out.

32 Lesser Bird-of-Paradise

SKELETON

Sexual Selection: Ornamentation

The example of intra-sexual selection of the fighting giant deer stags might imply that the males are the sex in control of who gets to mate. This is of course only part of the story. There are other mechanisms by which sexual selection can occur, including 'inter-sexual selection' or 'female choice'. While intra-sexual selection produces traits which help individuals outdo rivals of the same sex, inter-sexual selection favours traits in one sex (usually males) which attract the other.

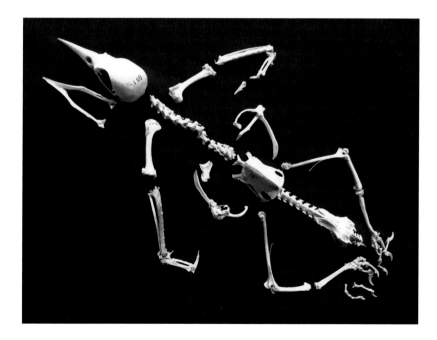

The cost of reproduction is rarely evenly shared between males and females, and this affects the way that each sex can go about seeking mates. Female birds-of-paradise do all of the heavy lifting when it comes to producing babies. First, they have to make the egg itself. Not only are bird eggs millions of times bigger than the sperm that males produce, but they are also comprised of highly energetically expensive protein (chiefly what makes up the egg white) and fats (chiefly in the yolk) to nourish the growing embryo. Female birds-of-paradise also do all of the nest-building. Once they've laid the eggs, the females do all of the incubating, and once hatched they collect all the food for their growing chick. After their 'encounter', all a male will do is fly off to try to find additional partners.

As sex comes very cheap to the males, the best evolutionary tactic for them is to mate with as many females as possible, producing a large number of young. The limit on their reproductive success is how many females choose them as mates.

Conversely, the limiting factor for a female's reproductive success is the significant time and energy she invests in producing and raising the young. This means that it benefits females to be very picky about their mate, because they're not going to have many of them. Their best tactic is to maximise the fitness of their offspring, ensuring that they survive and produce fit grandchildren. It is in situations like this, when investment in the young is really uneven, that sexual selection works most strongly. On the other hand, when the two sexes both invest heavily in bringing up the young (by gathering food for them, for example), then the payoff for reproducing more is decreased. In these cases the intensity of sexual selection is lower and males and females tend to look and behave more similarly. The way that sexual selection manifests in species like birds-of-paradise is to drive males to evolve characteristics that females will find attractive.

In this this way the females can select the 'best' males to be the father of the few babies they will raise, while the showy males can try and attract the highest number of females. As the total number of reproductive encounters has to be the same among males and females (one can't do it without the other), this obviously means that sexual success among males is uneven – some will attract many mates, some will attract few.

While females of these New Guinean birds are drab and brown, and generally bird-shaped (sensible naturally selected traits for keeping a low profile from predators), males have developed some of the most absurd colourings, ornamental plumes, dances and rituals of any animal (including shape-shifting dance-moves, outrageous sounds and cleaning their dance floors to perfection).

Sexual selection through female choice explains with how such odd features might evolve. It begins with females having a slight preference for an extravagant male trait. It might not have a selective advantage (the initial innate preference could be random – merely because it is attractive), but in some cases this could give the female information about the male's genetic fitness (for example, in order to grow and survive carrying around a giant peacock tail, the male must have avoided predators and significant parasites in his life). But as long as there is a genetic basis for the female finding the male's showy trait attractive, the scene is set for that trait to develop to be even more elaborate.

It's called runaway selection – females who find males with colourful plumage attractive will mate with such a male, and so produce young that have both the male's plumage and the female's inherited preference for that plumage. In time, the preference creates selective pressure for even more colourful plumage, and eventually we end up with birds-of-paradise.

The specimen photographed on p. 113 is a skeleton, and so doesn't show the fabulous plumage (perhaps it was a female, and so subject to the same biases as the collecting of giant deer in museums), however it does link nicely to this story. When Darwin published the initial explanation

of sexual selection in *On the Origin of Species* in 1859, the scientific community needed convincing of his ideas. This skeleton once belonged to the man most responsible for arguing on behalf of natural selection – Thomas Henry Huxley, one of the most influential biologists of the century. However Huxley was not overly supportive of the idea of sexual selection itself; there were some at the time who seemed to struggle to accept that female choice could have such profound influence on a male's traits.

Male lesser birds-of-paradise show off their extravagant plumage with elaborate dances to attract a mate.

33 Garden Snail

Sexual Selection In Hermaphrodites: Love Darts

So far I have discussed species in which sexual selection has impacted on both male and female members of a species, in terms of driving the evolution of weaponry and ornamentation in males (as well as accompanying behaviours), and the means to discern between them in females. What happens when a single individual can play both the male and female roles at once?

Two love darts used by snails to stab each other during sex, which increases the likelihood of their partner accepting sperm. These are about 1cm long.

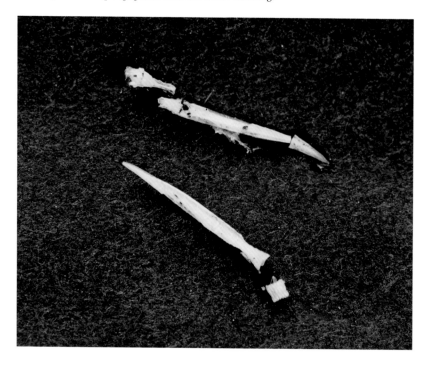

Darwin believed that simultaneous hermaphrodites – species in which all individuals carry both male and female sex organs at the same time – were immune from sexual selection, simply because an animal cannot be in an evolutionary conflict with itself. However, many examples of extreme behaviours and anatomical developments have since been found in the most overlooked hermaphroditic species like worms, slugs and snails.

Some hermaphrodites can self-fertilise (using their male organs to fertilise eggs in their own female organs), but mating between two individuals is common in others. In some species, during a given mating event one individual will play the male role and provide the sperm (the biological term is the sperm 'donor', which implies to me that the recipient is a grateful beneficiary, which is by no means always the case). In other species they simultaneously transfer sperm to each other. In others still they take turns in being the sperm donor.

In species where it is energetically 'cheaper' to play the male role, one would expect that sexual conflict would arise – how does one individual 'win' the battle, and convince the other to take sperm and fertilise their eggs with it? The same rules as before apply to hermaphrodites – the best evolutionary outcome for an individual playing the male is to father the *most* young, while the *fitness* of the young is normally the female's primary concern. In some free-living flatworms, the competition to be the male has reached actual battle stations – they carry out 'penis fencing'. Whoever wins the sword fight and successfully pierces their opponent's skin with their sharpened penis, and ejaculates, gets to be the father. The other flatworm has to bear the cost of reproduction (including a flesh wound).

Some horrific (from a human point of view) developments have evolved in gastropod molluscs – banana slugs chew off their partner's penis after mating (so they only get to mate once – at least as a male). Some species of garden slug have evolved penises many times the length of their bodies. Some produce giant packets of semen (which actually match the production of eggs in terms of energetic investment). But it is the common garden snail that I want to focus on – they stab each other with solid calcium-based swords to increase the reproductive success when playing the male role.

These pointed rods are called 'love-darts'. They appear in a number of snail species, some of which can reuse their darts to stab each other 3000 times during one mating. Garden snails only stab each other once, but the dart is designed to detach and stick in the mate (they can grow a new one in a few days).

The role of the love darts has been debated since they were discovered – mollusc researchers (malacologists) have suspected that sexual selection is the evolutionary cause, but it has not always been clear what the specific role of the dart was – is it a calcium gift to help the eggs grow? Does the mechanical stimulation of being stabbed increase the likelihood of an individual playing the female role (and if so why)?

It is now known that the dart is just a tool for delivering mucus which carries a hormone that alters what the other snail's female organs do with the sperm. Snails fertilise internally (the penis deposits sperm inside the vagina, and the sperm and egg meet inside the body). However, they can store their partners' sperm for up to 4 years after mating. They will have several mates in that time, potentially storing a bit of sperm from each of them. There is actually very little chance of a sperm ever reaching an egg, as the organ responsible for receiving sperm is attached to an organ responsible for digesting it.

During mating – which can last seven hours – garden snails cover their love darts in a mucus as it passes out of their body, and inject it by punching a hole in the other snail's skin. They then bring forth their penis and attempt to insert it into the other's vagina. Both snails do this at the same time, and each transfer a package filled with sperm.

Only a small number of sperm get to swim from the sperm-receiving organ to the sperm-storing organ – the rest will be digested. This means that there will be a strong selective pressure for snails to evolve means to increase the chance of their sperm avoiding digestion, being stored, and ultimately have a chance of fertilising an egg. Paternity is the evolutionary goal.

If it hits the target, the love dart delivers a bioactive substance to the interior of the recipient. This causes the recipient's female reproductive system to reconfigure – closing the tract that leads to the sperm-digesting organ. The result is that the number of sperm reaching the sperm storage organ is higher, and when it comes to fertilisation (potentially 4 years later), they father more young.

'Sperm donor' really doesn't sound like the most appropriate phrase.

34 Bed Bugs

PRESERVED SPECIMENS

Sexual Competition: Traumatic Insemination

The idea of being stabbed with a love-dart appears to be a rather violent outcome of the conflict that arises when the costs or benefits of reproduction are not equal between the sexes. However, I find some solace in knowing that, as hermaphrodites, both of the individual snails involved in the example above are stabbing each other. Yet, in some invertebrate species with two sexes, the evolutionary arms race has resulted in males piercing the skin or exoskeleton of the female and ejaculating sperm directly into her body cavity. This outcome of sexual selection (the evolution of traits which increase an individual's reproductive success, but not necessarily their survival) is called traumatic insemination.

Arthropods (the largest phylum of animals, which includes those with an exoskeleton and jointed limbs, such as insects, spiders, scorpions, crustaceans, centipedes and millipedes) don't have a circulatory system

like vertebrates. Instead of discrete blood vessels, their entire body cavity is simply filled with a fluid that comprises both the blood and the lymph (we bony creatures have separate systems for each). This fluid – called haemolymph – bathes all of the animal's organs. By piercing the body cavity and ejaculating directly into the haemolymph, the sperm injected by traumatic inseminators can simply swim to the females' ovaries, bypassing more 'traditional' routes through the reproductive tract.

While females of species which use traumatic insemination have evolved some counter-adaptations, it would appear that as it stands the males are gaining reproductive success at a cost to the female's survival as the mechanical harm done to her is often detrimental. Because males gain and females lose from this interaction, the scene is set for sexual conflict.

While female bed bugs do have a genital tract, it is not used during copulation. The males have evolved a sharp spine with which to deposit their sperm direct into the body cavity, piercing through the exoskeleton. Perhaps as an evolutionary attempt to reduce the physical harm done by such a wound, females have evolved tiny structures to guide the penis in through the wall of their abdomen less traumatically, but the wound is still severe.

Interestingly, in a study undertaken at Sheffield University by Alastair Stutt and Michael Siva-Jothy, the females didn't do anything to try and stop the males mounting and stabbing them. This could suggest that traumatic insemination has evolved as a means for the males to bypass any counter-measures the females had developed for controlling access to their genital tracts.

The study went on to find that multiple traumatic inseminations at a rate that occurs naturally does actually decrease the lifespan of the females, when compared to the minimum number of copulations they would need to maximise their fertility (males will mate with females more often than the females need them to in order to satisfy their own reproductive needs). The copulatory behaviour of the males is actually causing the females to die younger.

This doesn't come as a great surprise – the female is being stabbed in her abdomen with a sharp tool – it will be energetically expensive to keep healing those injuries, and also to fight off any infections that build up in the wounds. It doesn't help that several pathogenic bacteria were found on the males' penis itself. Spending energy recovering from the wounds from traumatic insemination means that they can't spend it on maintaining their health in other ways, and so they die younger.

The females appear to have evolved some mechanisms to reduce this harm, with structures both below and on their exoskeletons to guide the penis in as smoothly as possible, but it still seems like the males are currently 'winning' this particular evolutionary battle.

35 Seahorses

PRESERVED MALE AND FEMALE

Sexual Selection: Male Parental Care

When the two sexes within a species look different, they are said to be sexually dimorphic. Sexual dimorphism often results from sexual selection – the males have different evolutionary drivers from the females, and this means that their physical appearance is different. This can manifest in one sex being considerably bigger than the other; the development of weapons like males' antlers or horns in some species; or one sex displaying elaborate ornamentations such as colourful plumage or lengthy tails.

The pouch of the larger male seahorse is clearly visible. When 'pregnant', he can carry hundreds of tiny seahorses.

Seahorse males and females only really differ with respect to their sexual anatomy, but yet they are still considered an interesting example of sexual dimorphism. One sex has a brood pouch on their bellies for carrying their young until they are big enough to fend for themselves. It is the male.

So far in this collection of objects, the males have come across as the bad guys in stories of sexual conflict. This is, obviously, a moral judgement made through the human lens of right and wrong, not the impartial judge of evolution. In seahorses we have an example of a system in which females seem, to our anthropomorphic philosophy, to have the upper hand.

Males being significant or principal caregivers is not unheard of in the animal kingdom. There are 'good fathers' among plenty of birds, which can at least share the responsibility of incubating eggs on the nest and feeding the young. Among the amphibians, midwife toads are a good example of a species where the males guard the eggs (tied in a string around their legs), and there are a number of fish species in which the males guard eggs and fan them with oxygenated water.

However, when it comes to taking parental responsibility, male seahorses have gone so far as to effectively become 'pregnant'. They have a pouch on their fronts, with a tight opening through which the females will lay her eggs. In some species these can number in the hundreds. Once in the pouch the eggs are fertilised by the male releasing his sperm over them. Incredibly (for a fish), a placenta-like structure then forms to transfer nutrients to the growing embryos.

A heavily 'pregnant' male seahorse, with a very full brood-pouch.

It has been hypothesised that it is comparatively unusual for males to invest significant resources in raising young, as they can't always be sure that they are the true fathers of their mates' offspring. Unfaithfulness has been found to be very common in the animal kingdom, and evolution is unlikely to favour situations in which the male is duped into raising a rivals' young (even in humans, evolutionary biologists have suggested that babies look more like their fathers when they are first born for this reason – to reassure them that they are theirs). Statistical relationships have been suggested that demonstrate that in species where males have more certainty that they are the father there tends to be greater paternal investment.

This is why the seahorse's circumstance is so interesting – because he fertilises the eggs inside his own body, a male can be very sure that he is the father of all the eggs. This puts him in a similar boat to the females of other species, where they are the primary investors in making babies.

Did the males' taking over all the costs of raising young (aside from producing the eggs themselves, of course) evolve as a means to convince the female to mate with them? Is it an extreme example of those situations in which males seduce females by providing them with 'nuptial gifts' – energetically rich food parcels to help them nourish the eggs, or demonstrate what good hunters they are?

There are examples among pipefishes, which, like their close relatives the seahorses, have male parental 'pregnancies'. Here the sexual role reversal has gone even further, however. Some pipefishes are much more sexually dimorphic than seahorses – they differ in size and even bright colouring. However, in these cases it is the *females* which are more strongly decorated, much like our male birds-of-paradise. What's fascinating about this is that these species are polyandrous – the females have multiple male partners, while the males tend to have just one. Seahorses, on the other hand, tend to be monogamous and are much less dimorphic (aside from the male brood pouch).

Where reproductive success is strongly unequal, sexual selection has driven sexual dimorphism, and in the case of pipefish the males are in the position to be choosy about whom they mate with. It's a perfect mirror opposite of the situation in species like deer and birds-of-paradise, where the males aim to have multiple female partners, and the females are choosy. It is true sex role reversal.

36 Gorillas

MALE AND FEMALE SKULLS

Sexual Selection: Size

Size is one of the most obvious ways in which the males and females of a species can be different from each other. In many species, particularly among birds, insects, frogs and snakes, the females are considerably larger because of the energetic and physical requirements of producing and carrying eggs. However, when the evolution of an animal's size is affected by competition for mates, sexual selection tends to drive the males to be bigger.

Among the primates, gorillas are one of the most sexually dimorphic species. The males can reach 180kg, while the females are only approaching half as big. Gorillas live in small groups of five to ten individuals, containing a few (normally unrelated) adult females, their offspring and a single huge adult male – the silverback – who is the father to all the young. As they mature, both males and females leave the group they were born into. At that age, the adolescent sons have little chance of outcompeting

The larger teeth and the solid ridges of bone on the male gorilla's skull are sexually selected traits.

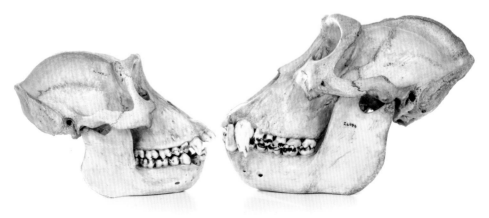

their father to gain leadership of the group and the mating benefits that come with that position. Once they come of age they leave to spend several years wandering the forest alone, until they are big enough to attract their own females or beat an established silverback in a fight and take over his harem. Females leave the group they are born into once they are ready to breed, because their only option there would be to mate with their father.

This system means that there is a huge difference in how successful a male can be when it comes to reproduction – those that can protect a harem for a long time will have many young, and those that can't won't. This puts sexual selection to work. Another factor is that the silverbacks protect the females and young members of the group from predators, as well as from other passing males who may want to take over. When a new male challenges a group's silverback and wins control of the harem, he tends to kill all the juvenile members of the group. Without these young to take care of, the females' reproductive systems will boot up again (species which can only become pregnant by mating when they are on heat stop being reproductively receptive while they are caring for dependent young), and they become ready to conceive the new male's babies. If a resident male can fight off a challenger, he is protecting his offspring from a brutal death, and thereby his own genes.

As a result of these two related pressures – to protect his group from predators and to be strong enough to beat another gorilla in a fight for breeding rights with the groups' females and the lives of his young – male gorillas have grown to be huge.

Looking across the primates, the size discrepancy between the males and females is a good indicator of how mating is arranged in that species. Comparisons have shown that the more females there are for each male in the breeding group, the greater the difference in size. So in species like gorillas where one male, and one male alone, breeds with all the females in his group, the sexes are the most dimorphic. In species which are monogamous the sexes are much more similar, as the males don't need to come to blows in order to gain the right to father young.

When you look at this pair of gorilla skulls, aside from their size the other significant difference is with their teeth – males have much larger teeth, including massive canines. It may be obvious to say that bigger bodies require bigger teeth (indeed, bigger everything) in order to process enough food to sustain their bulk. The massive ridges of bone on the skull are linked to the size of the teeth; they carry the jaw muscles. However, male gorilla teeth are bigger than they need to be to account for the difference in size compared to the females.

Differences in primate teeth between the sexes are another good indicator of the system they have for breeding. After the difference in body size is taken into account, males have bigger teeth than females

A large male silverback gorilla (*left*) and a female. The female is significantly smaller and lacks the huge jaw muscles and domed skull.

in harem-forming species like a gorilla, while monogamous species are roughly equal, dentally speaking. This is also a sign of sexual selection – the teeth and the massive muscular apparatus that go with them are a significant part of the intimidation routine that males go through prior to battle. Bearing their teeth, roaring and generally making themselves look huge are key tactics in convincing the other guy to back off.

At the other end of the body, there is another way of predicting how a species breeds, and that's the size of their testes. Gorillas have some of the smallest testes of any primate species, after taking body size into account (obviously a 180kg gorilla's testes are bigger in absolute terms than a 200g marmoset's). This is because the females only mate with one male, so he only needs to produce as much sperm as is necessary to fertilise an egg, rather than to flood out the competitors' sperm. This isn't the case with all species.

Sperm competition is a very important factor in sexual selection. It occurs when females mate with more than one male, and so not only does he have to compete to get the chance to mate with her, once he does, there is evolutionary pressure to come up with ways to increase his chance of paternity. There are many species that go to some lengths to physically remove a previous male's sperm from the female's body. It has been suggested that the shape of the human penis is an adaptation to dislodge previously 'deposited' sperm from the vagina, and that sexual selection favours longer penises in order to release the sperm closer to the egg, to increase chance of being the one to fertilise.

A solution to the question of sperm competition is to produce more sperm – by effectively flooding the system you can increase your odds of gaining paternity. Larger testes are required to produce more sperm, and this is exactly what we see in species where multiple males each mate with reproductively receptive females. Chimpanzees have this system, and they have the largest testes of any primate: a 45kg chimp's testes weigh around 120g. A huge gorilla's by comparison, only weigh 30g.

In case you are curious, the relative weight of a human's testes compares more favourably with species that are either monogamous or have a single-male breeding system (like a gorilla), than that of a chimp.

37 Walrus

Secondary Adaptations: Tusks

When looking at animals that appear to have huge weapons, it's very easy to assume that sexual selection is at play and they have evolved to help individuals to be a successful breeder by fighting off rivals. That may well be the case, but when the 'weapons' are found in both the male and females, we should look for alternative or additional explanations – straightforward natural selection may be at work.

Both male and female walruses have sizeable tusks. These modified canine teeth can reach a metre in length, which is pretty astonishing when you consider that males do use them in battles over females, physically stabbing their rivals with these spear-like teeth. The skin of adult males is heavily scarred from such battle wounds – fortunately they have tough skin and a very thick layer of blubber under their skin which not only protects them from the Arctic cold, but presumably goes some way in reducing serious injury after being tusked.

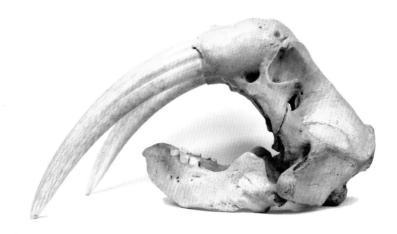

A walrus feeding along the sea floor.

As with all comparable appendages like antlers and horns, walrus tusks can be used merely as a threat, avoiding physical blows. Establishing dominance by showing off their tusks is seen in both male and female walruses, between members of the same sex. However, that isn't the whole story.

Quite often anatomical features evolved for one purpose (like demonstrating a pecking order) but are then found to be rather useful for something else. This is called a secondary adaptation and walrus tusks have been found to be very useful for a number of non-combative behaviours.

Because males and females both have them, biologists have long been searching for explanations for the walrus' tusks aside from same-sex competition (although this may play a role too). These giant seal-relatives eat a huge amount of bivalve molluscs (chiefly clams), and other invertebrates that live in the mud at the bottom of the sea. For a long time it was assumed that walruses use their tusks to rake their prey up out of the silt. However, by analysing the minute scratches on the surface of the tusks, this possibility was ruled out. Because they point downwards past their chins, to use them as a rake they would need to dig them into the mud with a backwards and downwards motion. This would cause the tusks to be scratched on the inside surface nearest the body, but no such scratches were found.

Instead, the scratches were on the outermost surface where they curve outwards, in a direction parallel with the body. This suggests that the tusks are dragged flat through the mud as the walrus swims forward, which would not help with raking.

Another unusual discovery was made by looking at the wear and tear of the peg-like teeth at the back of the walrus' mouth. Despite the tens of kilos of clams that walruses eat each day, it is extremely rare to find clam shells in a walrus' stomach: they only swallow the fleshy parts of the mollusc. However, the cheek teeth behind the tusks show no evidence of chewing shells. So if they aren't smashing the shells open with their teeth, how are they getting the meat out of the clams?

The answer to this question demonstrates some of my favourite evolutionary adaptations in any skull, and it explains how a 1,800kg giant can deftly eat a tiny clam without using its hands (they have highly modified flippers for swimming): they *suck* the meat out. This requires a lot of force, and a number of special anatomical features.

First, to get the shell in position. The muscles of the snout and the beard of short, stiff whiskers are more highly developed in the walrus than in any other carnivore. The whiskers actually function like hands to manipulate the prey in the mud, selecting live clams from other objects in the murky seabed. The front of the walrus' skull is perfectly shaped to hold the chosen clam's shell in its pursed tip.

The lower jaw of all mammals is formed of two separate bones – one mandible on each side. The extent to which these two bones are fused together varies between species, but in most carnivores it is relatively loose. The walrus, on the other hand, has an absolutely solid lower jaw – it can withstand a lot of pressure without changing shape at all.

The bones on the inside of a walrus' mouth are arranged so that the palate is highly arched, forming a tube. The tongue is more or less cylindrical, fitting closely inside the tube and operates like a piston or a plunger. When the clam is in position at the front of the mouth with its shell locked in by the solid mandibles, the tongue pumps backwards and a vacuum is formed. This sucks the meat of the clam straight out, leaving the hard parts to be discarded intact.

This means that the tusks are not used in feeding at all. However, a number of different uses have been observed to demonstrate what walrus tusks *are* secondarily adapted for, aside from fighting. One explains where the walrus' scientific name *Odobenus* comes from: *odous* is Greek for tooth, and *baino* is Greek for walk. They are 'tooth-walkers' – walruses use their tusks to climb out of the sea onto the ice, as if they were ice axes. This could help explain why males' tusks are larger than females': as males weigh more, they need stronger tusks to bear their weight.

Walruses have also been observed using their tusks to anchor themselves into an ice floe as they sleep or rest in the water. By hooking their tusks over the top of the ice, almost like grappling hooks, they can keep their heads above water and breathe safely. Tusks are also used as ice saws to widen or create holes through which the walrus surfaces to breathe, and to climb in and out of them. Despite only being a few centimetres in diameter, walrus tusks can cut through surprisingly thick ice.

The lesson here is that it could be easy to assume part of an animal has evolved for a single purpose, but in many cases there are several secondary adaptations that have arisen to fulfil completely unrelated needs.

38 Dugong

SKELETON WITH VESTIGIAL PELVIS

Family Ties: Homology

For me, there is no greater evidence for evolution than the dugong's pelvis.

The science of taxonomy – putting species into groups based on their evolutionary relationships – depends on how many characteristics species share. If they share a lot of features – be they anatomical or genetic – they are assumed to be closely related and belong in the same group. If they share few, they are from different groups. Humans share few characteristics with a slug: we are distantly related. Humans share many features with a dolphin: we are more closely related.

The challenge is deciding whether two species really do share a given characteristic. We need to determine whether they both inherited the same feature from their shared ancestor (such a feature is said to be homologous), or whether a similar thing evolved twice in two lineages (the feature is analogous).

I'll return to analogous characteristics with the next object, but the most widely used example of homology is the limb of all land-living vertebrates. Our shared ancestor had an unspecialised limb with one long bone at the top, two long bones at the bottom, a collection of small bones at the wrist or ankle, and then five digits. This basic template has been modified

The rods of bone behind the dugong's ribs are the evolutionary remnants of its pelvis – evidence that they evolved from animals that lived on land.

through time to become the anatomically distinct wings of birds, bats and extinct pterosaurs; the hooves of cows and horses; the flippers of whales and seals; our own dextrous hands, and many other specialised adaptations. Over the course of their evolution, some of the groups have lost digits, some gained them, some lengthened parts of the limb, and some have fused the component bones together. The key thing is that they are all evolved from the same ancestral body part – they are homologous.

Whether or not they have since been modified, homologous characteristics tell us that two species shared a common ancestor, which also had the same feature.

How do we know that the wing of a bird has the same ancestral route as our own hand? There are at least three lines of evidence used to investigate whether two adaptations are homologous: morphology, genetics and the fossil record.

We would expect homologous features to be very morphologically similar – each bone articulates with the same neighbouring bones, the same muscles attach to them, and the same nerves and blood vessels serve them, for example. We would expect the same genes to control their development and function. Through geological time, we would expect to be able trace the changes from the ancestral form through the fossil record.

Not all lines of evidence will be available in all situations, as DNA doesn't preserve well in fossils, for example, or a body part might be so reduced that the way it relates to the rest of the body can no longer be determined, but these are all tools in judging whether two species share a common ancestor.

Dugongs are a large species of herbivorous marine mammal from the Indian and Pacific Oceans, and are sometimes called sea cows. The only other living members of their order, the sirenians, are the manatees. This name came from the aquatic sirens of Greek mythology, as sailors apparently confused them with mermaids in the early days of exploration (though I would think you would need to be at sea for a very long time to confuse the front end of a manatee with a female human).

As adaptations to their fully aquatic lifestyles, dugongs and manatees have completely lost their hind limbs and modified their forelimbs into paddle-like flippers, much like whales and dolphins (which are only very distantly related). They swim by pushing their tails up and down to propel themselves forward. But despite the fact that they have no legs – no thigh bone, no calf and no ankle – they do have a pelvis. Two small vestigial rods of bone are visible, suspended under the spine towards the back of the dugong's skeleton. Evolution has shrunk the legs to nothing, but the pelvis remains. Perhaps this is because natural selection has no real need to lose it – it is embedded deep within the muscle tissue, and so does not cause drag in a way that unnecessary limb bones would. Or

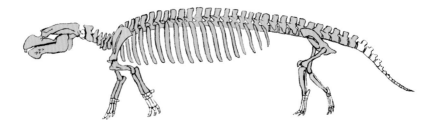

The 50-million-year-old sirenian *Pezosiren*, which had four legs.

perhaps it is still shrinking and in a few million years' time it will disappear completely.

This bone is proof of evolution, as why else would an animal with no legs have a non-functioning pelvis unless it evolved from an animal that did have legs?

We know that the dugong's pelvis is homologous with other mammal pelvises for a number of reasons. Based on similarities within both their molecular make-ups and their skeletal anatomies we now know that sirenians belong in a group with elephants. Together they form the Tethytheria. There is now also compelling fossil evidence of how the sirenian pelvis started off.

In 2001 a 50-million-year-old fossil called *Pezosiren* was described from Jamaica. It is the most primitive member of the dugong's group – very much recognisable from the shape of the skull, but it had four legs. Unlike the dugong's pelvis today, *Pezosiren*'s pelvis was solidly attached to the spine, which means its legs could bear the weight of the animal on land, but many other features of the skeleton show that it spent most of its life in water, like a hippo. It represents a clear step on the evolutionary route to the appearance of dugongs.

39 Box Jellyfish

PRESERVED SPECIMEN

The Same But Different: Analogy

The counterpart to homology is analogy. While homologous features are similar in two different species because they both descended from a common ancestor which also had the feature, analogous features have arisen independently in two different lineages.

Analogous features can cause taxonomists trouble, as when two species have characteristics in common it can either mean that they belong in the same group (because they are homologous, with the same evolutionary history) or that the two groups have separately come up with similar adaptions to solve a similar problem. These analogous features don't help taxonomists group the species together, and they have to be careful deciding whether the shared feature arose once in a common ancestor, and was passed down to the two related groups, or whether they arose twice: once in each of the species' separate evolutionary histories.

A simple example of analogy is legs – both giraffes and beetles have legs, but they have evolved them independently of each other. If we trace their evolutionary history a very long way back in time, to some of the earlier phases of animal evolution, we know that the last common ancestor of both giraffes and beetles did not have legs – it was likely a worm-like blob living over 530 million years ago. In the intervening time, the groups that led to giraffes and beetles both evolved legs, but they don't tell us anything about how giraffes are related to beetles: they are analogous. We know this because although the function is the same (walking), the anatomy, fossil record and genetics of the two kinds of leg are completely different.

An even more distant relative of the giraffe (and indeed of all of vertebrates) is the box jellyfish. Jellyfishes belong to one of the oldest groups of animals – the Cnidaria – whose fossil record may date back around 600 million years (chapter 3). The only living animal groups that pre-date Cnidaria are sponges, comb jellyfish and the microscopic placozoans. Aside from some of the general features that all animals share, there are few homologous characteristics that unite jellyfishes and vertebrates.

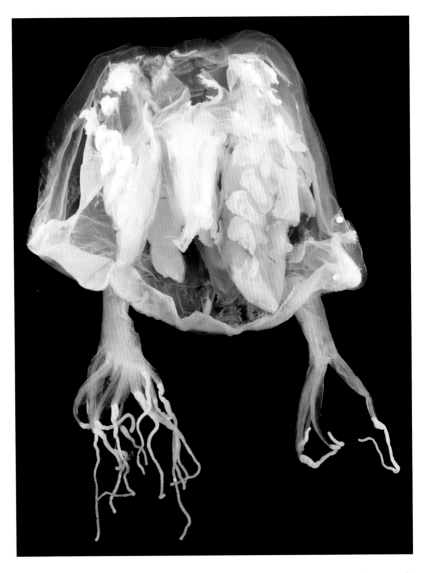

The eyes on each 'side' of box jellyfish evolved independently from the eyes in other animal groups.

One of the most remarkable features of box jellyfishes is that they have complex eyes. As they have evolved independently in this group and in vertebrates, they are analogous. This is also called convergent evolution.

Complex eyes might seem like pretty remarkable organs, and indeed they are, but they are not unusual. Image-forming eyes with lenses have arisen independently in several animal groups: aside from vertebrates and box jellyfish, there are velvet worms, bristle worms, arthropods and molluscs with eyes. In fact, eyes have evolved separately in a number of mollusc groups at different times. Despite their apparent complexity, eyes are such beneficial adaptations that we shouldn't be surprised that they are so common. Natural selection would exert a strong pressure to develop them from simple light-detecting cells.

Most jellyfishes are rather simple creatures: they don't have a brain; just a loose network of neurons control their functions. They drift aimlessly in the sea, going with the flow. However, one group, the box jellyfishes, or cubozoans (literally 'cubic animals'), have developed eyes and tiny concentrations of nerve cells around them to process the visual information they receive. The eyes are located near the centre of each of the flat sides of their box-shaped bells. They actually have six eyes on each side (so twenty-four in total) – four are very simple receptors that basically just tell light from dark, but the other two have lenses and can form images. These sit on a little protruding stalk, which has a solid crystal called a statolith near its base. Not only does the pull of gravity on the statolith inform the animal which way up it is, it also acts as a weight, ensuring one of the lensed eyes always points towards the surface, whatever the body's orientation (think of a sock with a floating ball on one end, and a heavy ball on the other, no matter how you pull the sock through the water, the end with the floating ball will always be above the heavy ball).

A 2011 study by Anders Garm, Magnus Oskarsson and Dan-Eric Nilsson at the universities of Copenhagen and Lund found that box jellyfishes use their eyes to seek out their favoured habitat. Unlike other jellyfishes, cubozoans can actively swim in a given direction to suit their needs. They hunt little crustaceans called copepods swimming near the edge of mangroves. Some box jellyfishes are famous for being among the most venomous creatures on earth, carrying stings that can kill humans. It's rather amazing that such powerful toxins are required to catch their tiny crustacean prey. Hawksbill turtles, the chief predator of box jellyfishes, are immune to their sting. (I can attest to this, as I was recently swimming – perhaps foolishly – in some mangroves off northwest Australia when a turtle surfaced 3m away with a box jellyfish in its mouth. I like to think it saved my life.)

The study showed that the jellies use their image-forming eyes to ensure that they stayed within sight of the mangrove tree canopy, where their prey is found. If the canopy approached the edge of their vision, they swam towards the trees. This is beneficial as if the trees go completely out of view they risk wandering into the open ocean, where they may starve.

This shows that animals don't need a true brain to exhibit complex behaviour, and it's also a fantastic example of convergent evolution – where an *analogous* adaptation (in this case, eyes with lenses) – arises independently in separate taxonomic groups.

40 Striped Possum

SKULL

Convergent Evolution: Wood-Pecking And Politics

Zoology is tribal. To the outside world natural historians present a united front: the geologist is my brother and the botanist is my friend. But hidden within are genial rivalries. You might find that those noble folk studying the less celebrated animal groups carry a certain disdain for the glamourous animal fanciers. In palaeontology, fossil coral experts cry themselves to sleep at night when yet another dinosaur story makes the newspapers. In zoology, my entomological friends are constantly moaning about how much limelight the mammals get, when their insects are infinitely more numerous and diverse. However, there is nothing more mainstream than primatology – primates hog the stage and perhaps stop other species getting the attention they deserve.

The marsupial striped possum has evolved features that are extremely similar to the aye-aye, a primate.

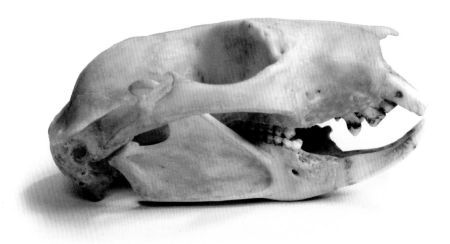

My own passion is mammals, so I would certainly be considered on the mass-popularity end of the spectrum, but here I present an un-famous species lost in the shadow cast by a much-celebrated primate in a similar ecological niche.

Striped possums are the marsupial version of aye-ayes, the odd-looking Madagascan primates that are famed for their ability to locate and remove insects boring through tree trunks and branches. Both species have convergently evolved a long, spindly finger (the possum's genus *Dactylopsila* means 'naked finger'), and protruding, gouging incisors. They both use their teeth to lever open a hole in bark and wood, and then use the long finger to hook out their beetle larva prey. They live and feed in trees. All of these shared adaptations arose independently in the two groups and this makes them my favourite example of convergent evolution.

This skull belongs to *Dactylopsila trivirgata* and was collected in New Guinea – the home of four species in the striped possum family. Interestingly, this species (but not the other three) also extends down into the rainforests of far north-eastern Australia, which could suggest that the population there is a recent colonisation from New Guinea, but a long-term fossil record of striped possums in Australia dates back around 25 million years.

If striped possums and aye-ayes both have an incredible suite of adaptations to catch wood-boring beetles, why are aye-ayes so famous, when so few people have heard of striped possums? Even in their native Australia you would be lucky to find a resident of Sydney or Melbourne who recognised them.

There are three ways to be a mammal, depending on how they develop their young: egg-laying, marsupial or placental. We are placental mammals, along with the vast majority of mammalian species (there are more than 5,100 species of placental). This group give birth to well-developed young after a long period of growth in the womb. The development is then finished off with a relatively brief period of suckling milk on the teat. The approximately 335 species of marsupial do the opposite: they give birth to tiny jellybean-like babies after a very short period in the womb. Most of the babies' development is done while continuously suckling milk on the teat, often in a pouch. Finally, there are just five species of living mammals that lay eggs – the platypus and echidnas.

Marsupials are probably the most downtrodden and ignored of the mammalian kind. They are often described as being inferior, evolutionarily limited and 'less advanced' than placental mammals. Today, except for one American species (the Virginia opossum) and a couple in Mexico, they are only found in Central and South America and Australasia, although their fossil remains have been uncovered from every continent.

There are convincing suggestions that these inaccurate denigrations of marsupials as 'inferior' is or was a colonialist political mindset born of the

Both aye-ayes (*left*) and striped possums (*above*) have evolved an elongated middle finger to extract grubs from holes they cut in tree trunks with their front teeth.

implication that Australia and the southern continents are less advanced. Looking back to the earliest days of European invasion, and even the accounts of Captain Cook and his crew from their landings on the east coast in the 1770s, there are constant and outrageous undertones that life in Australia, both faunally and sociopolitically, was somehow lower-grade than in the northern hemisphere.

Perhaps this is why marsupials are so often overlooked and dismissed. Add in the fact that aye-ayes are primates (like us humans) and we might understand why all the mammalian wood-pecking glory is given to aye-ayes instead of striped possums.

Striped possums have one of the largest brain to body size ratios of any marsupial. Marsupials have often been dismissed as second-rate mammals as people thought that they had smaller brains than similarly-sized placental mammals. It was believed that for a large brain to develop a long period of development in the womb was needed. Considering how easy this theory should have been to dis-prove, it is incredible that it took until 2010 for zoologists to properly check. Colleagues of mine at University College London did just that. Vera Weisbecker and Anjali Goswami plotted brain size against body size for both marsupials and placentals. They found the comparison was significantly affected by the inclusion of one abnormally large-brained group of placentals – the primates, and this skewed the data. When they took primates out of the comparison, no difference was found between the brain size of marsupials and placentals with the same body size (in fact, for smaller species, marsupials had larger brains than similarly sized placentals). Why did it take over 240 years of marsupial zoology to right that injustice? This is just another example of how primates have been allowed to take the limelight away from marsupials.

41 Slow Worm

PRESERVED SPECIMEN

Convergent Evolution: Getting Legless

Not all legless reptiles are snakes. The slow worm is just one of the many kinds of legless lizards. In the United Kingdom there are three native species of lizard (and three native snakes), but one of them has no legs – the slow worm. It's probably unnecessary to say that slow worms are neither particularly slow (especially when compared to actual worms), nor are they worms – one of the many quirks of English common names.

The complete or near loss of limbs has evolved in lizards a great number of times – it is a fantastic example of convergent evolution. Some entire lizard families are limbless. Other families contain a few species with tiny vestigial limbs, while the rest are limbless. Some families are mostly 'normal' four-limbed species, with limblessness, near limblessness, or two-leggedness having evolved in certain lineages independently. The biggest lizard family, the skinks (of which there are 1,500 mostly leggy species), includes numerous groups that have at separate, multiple times lost their limbs in Africa, Europe and Australia. In most cases of legless lizards, some remnants of the hindlimbs are visible, often by the presence of scaly flaps. This is also true of boa constrictors – this relatively primitive group of snakes has retained tiny vestiges of their ancestors' hindlimbs.

Leglessness evolves when the legs become a hindrance rather than a help in an animals' locomotion, and in lizards this is normally to do with burrowing or moving through leaf litter. Essentially lizards have found that it is more effective to 'swim' through the soil and other obstacles, pushing their way through little gaps with their heads. This makes sense because lizards' arms aren't that close to their snouts, so using them to dig can be a bit awkward. This is also one of the main hypotheses for how and why snakes evolved, although there is some evidence to suggest that swimming might also favour losing limbs.

Despite the superficial similarities that have arisen from independently evolving leglessness, there are a number of ways to tell snakes and legless lizards apart. These are good reminders that convergent evolution can always be spotted by looking at the anatomical details of the animals involved. Natural selection drove the groups to develop the same *function* for locomotion without legs, not necessarily the same *form* in which they do it.

First, lizards usually have holes for their ears, even if they are slightly covered by flaps. Snakes do not. Second, most lizards have eyelids (they can close their eyes), while snakes do not – instead they have a clear scale 'spectacle' over their eyes. You can also spot this in a snakeskin that has been shed – there is no hole for the eye. Third, many lizards have a fleshy tongue. (Monitor lizards are among those that don't – they have long thick forked tongues, and may be the group from which snakes evolved.) Snakes, famously, have long, thin forked tongues. Legless geckos can be tricky as they don't have eyelids, but the tongue confirms that they are lizards as they use their fleshy tongues to clean their eyes with, which snakes never do.

The next thing to look for when attempting to discern a snake from a lizard is whether it has a long tail. This may sound like a stupid question – aren't snakes all tail? Actually, the tail is the bit of the spine behind where an animal's pelvis is, or would have been. This can also be determined by where the animal's urogenitary opening is – the 'vent' is a scale-covered slit through which reptiles do their excreting, defecating and reproducing. Lizards tend to have long tails: their vents are a long way from the tip of their tails. To complicate things, many legless lizards (such as legless geckos and skinks) can shed their tails as a predator-distraction trick, making their tails look shorter, but the presence of a re-growing, stumpy tail is also evidence that it's not a snake. Conversely snakes have short tails: their vents are a long way from their heads (and snakes can't drop their tails).

Finally, the belly scales tend to be different in snakes and lizards. Most lizards have two or more rows of belly scales whereas most snakes have just one row of wide scales (the 'worm-like' blind snakes and file snakes

Legless western slender glass lizards are superficially snake-like, but have external ear openings (visible past the corner of its mouth) and eyelids, which snakes lack.

are exceptions). Otherwise the scales on a lizard's back look much like the scales on its belly, whereas snakes have very different sets of scales. Snakes move by undulating elongated rectangular scales (or sometime pairs of scales) in a wave formation down their bellies. In turn, each one grips and pushes against irregularities on the ground. This means they can't move over Teflon.

I should emphasise that there are exceptions among lizard-kind where they share one or two of the snake-like features, but presence of just one of the typical lizard characteristics should be enough to tell you that a given specimen is not a snake.

To add a taxonomic note of complication, snakes actually evolved from a group of lizards (an alternative theory is that snakes evolved from the marine mosasaurs, but they themselves probably evolved from within the lizards). This means that, by the rules of taxonomy the group 'lizard' has to include snakes, as valid taxonomic terms have to include *all* descendants of the group's common ancestor. So technically all snakes are just another kind of legless lizards. However that isn't regarded as a particularly useful way of looking at things, as all snakes are more closely related to each other than they are to (other) living legless lizards, as all snakes have a single common ancestor. By contrast legless lizards are not all closely related to each other – they do not form a distinct taxonomic group as leglessness evolved many times among the lizards.

42 Narrow-Bordered Bee Hawkmoth

Playing Nasty: Batesian Mimicry

Convergent evolution is one explanation for when two species appear similar but are not closely related. It is the result of the two species acquiring similar characteristics to exploit the same niche. However, not all animals that have evolved to look alike are the outcomes of convergent evolution. Sometimes mimicry is at work.

Despite appearances, it floats like a hawkmoth but does not sting like a bee.

There are a number of different kinds of mimicry (see the next chapter), but perhaps the most well-known is Batesian mimicry. This involves a harmless species 'tricking' a predator into thinking it is harmful, resulting in it being left alone. It is named after Henry Walter Bates (1825–92), a colleague of Alfred Russel Wallace – the co-discoverer of natural selection (a man who is regularly forgotten in the shadow of Charles Darwin). Bates and Wallace spent time together in the Amazon, where during his eleven-year expedition there Bates noticed that certain species of butterfly had almost indistinguishable wing patterns from other species, yet were not related.

As a defence against predators, very many species have evolved chemical protection which makes them either dangerous to tackle – through venoms and poisons – or simply taste awful. It is a common evolutionary tactic seen in many diverse groups including bees, fishes, frogs, sea urchins, millipedes and beetles. It sometimes isn't enough to simply *be* toxic or distasteful – it is also beneficial to let the predators know that you are toxic or distasteful. Once an individual has been captured by a predator, they may be killed before the predator finds out about their defence mechanism, which defeats the purpose.

It is very common for species that are venomous, poisonous or distasteful to warn potential predators not to eat them. This is most readily observed in species with warning colouration – bright, conspicuous markings which show clearly that the individual is best left alone. This is called aposematic colouration. Bees and wasps are a classic example: red, yellow and black markings are aposematic signals to tell predators that they carry a sting.

Ladybirds' red and black wing cases are aposematic warnings, advertising that they taste genuinely terrible. They aren't toxic, but when threatened they exude a thick yellow liquid called 'defensive fluid' – actually a derivative of their haemolymph (the insect equivalent of blood) – out of joints in their exoskeleton. It's quite common to find it on the palm of your hand after a ladybird has taken flight. People who are curious enough to taste these little liquid drops rarely do it a second time. It is truly disgusting – this is why these beetles are so brightly coloured.

Natural selection has driven unpalatable prey species to develop aposematic colour signals alongside toxins or distastefulness as a means to affect the predators' behaviour *before* it strikes. Natural selection has also driven predators to learn to *avoid* brightly coloured prey items to stop them being stung, poisoned or simply disgusted. The warning and the response to the warning go hand in hand.

The presence of a behaviour in predators to avoid certain colours and patterns provides an opportunity for non-toxic species to exploit. By evolving similar patterns to a toxic or distasteful species, they can enjoy the benefits of not being eaten whilst avoiding the energetically expensive trouble of actually developing the toxin. This is exactly what was happening in Bates' Amazonian butterflies: species that were perfectly edible had evolved to look almost identical to the aposematically marked foul-tasting species.

Bates' butterflies are incredible – different butterfly species evolving often indistinguishable combinations of coloured stripes and spots to look identical. There are a number of 'pairs' of matching species – one unpalatable to predators (the 'model') and one not (the mimic). However, the narrow-bordered bee hawkmoth is even more impressive. This is a moth that has evolved to mimic a bumblebee – two very different groups of insect.

Bee hawkmoths have transparent wings and a furry abdomen striped with black and orange. I have shown the specimen on p. 142 to hundreds of museum visitors and I don't think a single one has identified it as anything other than a bee, and that's with a dead specimen held in their hands. They can only be harder to distinguish when they are alive and flying around. At first glance, it certainly looks bee-like, but if you look with a critical eye the thicker and longer antennae, the curled proboscis and tiny narrow head do give it away as a moth.

Batesian mimicry has one potential flaw – if predators eat a brightly coloured but edible *mimic* (rather than the actually toxic or unpalatable model) and discover that it *isn't* poisonous or distasteful, the predator may fail to learn to avoid the warning colours. The presence of mimics weakens the power of the warning signal in the model, which is bad news for both of them.

As a result, mimics are generally less abundant than their models, making it more likely for the predators to learn to avoid the warning signs (as the odds of predators encountering a model are greater if they are more common). This is an example of frequency-dependent selection – when the value of an adaptation is affected by how common it is in a population. The more common the mimicry, the less powerfully it will be selected for by evolution.

It's actually quite hard to pinpoint exactly which bumblebee species narrow-bordered bee hawkmoths are mimicking. This may be because the potential predators haven't evolved the ability to discern, or it could be to do with frequency-dependent selection. By looking generally 'bumblebee-like' they might be widening the ratio of models to mimics. This means they could be confused with a number of bumblebee species – not just one – thereby increasing the possible number of models, and decreasing the chance of a predator un-learning its avoidance behaviour.

43 *Heliconius* Butterflies

PINNED SPECIMENS

Gang Colours: Müllerian Mimicry

Along with the bumblebee mimics, there are some other mind-blowing mimics among butterflies and moths. The finely scaled wings of this group are clearly highly adaptable for the evolution of an extremely diverse range of patterns and colours. But it's not just the wings that are used in mimicry – the shape and colour of the body and legs (as we saw in the last chapter) are also put to use in these deceptions.

There are moths in the family Sesiidae which are almost indistinguishable from wasps, encouraging predators to avoid them through Batesian mimicry. They have shiny abdomens with tight black and yellow stripes, sharply pointed abdomens suggesting a sting, and transparent, veined wings. Some even have markings that make them look like they have the characteristic narrow 'waist' of a wasp.

It's not just other insects that moths can mimic. In South America there is a species of moth called *Phobetron hipparchia* that appears to mimic poison dart frogs, which are brightly coloured to warn predators of their deadly poisonous skin. Incredibly this is actually a double-mimicking species, as its caterpillar has huge fleshy outgrowths along its body to resemble legs, and a fuzzy brown colouring which makes it look like a tarantula. Its common name is charmingly the monkey slug.

Another South American moth *Trichophassus giganteus* seems to have a similar evolutionary tactic. Its head and thorax are brown and extremely furry, as are its first two pairs of legs which are held forwards – its resemblance to a tarantula is uncanny. Interestingly, however, its wings and abdomen are deeply camouflaged to blend in with dead leaves and bark. Perhaps this is to help uphold the tarantula mimicry, as predators might not be fooled by a spider with obvious wings.

The species on the right are all distasteful to predators: they are Müllerian mimics of each other. Those on the left are edible – each is a Batesian mimic of its neighbour. By precisely mimicking their warning colours, they benefit from predators avoiding them.

Not all moth mimicry is visual – the death's-head hawkmoth (a relative of the bee hawkmoths, made famous by *Silence of the Lambs*) appears to mimic bees in different ways. They primarily feed on honey, stolen from within bee hives. Normally this would be a very dangerous place for a non-bee; however, death's-heads have evolved a chemical signal that mimics the 'smell' of bees. The can trick bees into leaving them alone while they plunder the honey. Not only that, but they may be auditory mimics as well. When threatened these large moths can emit a screaming sound by forcing air through their proboscis, scaring away potential predators. It has been suggested that this may also be used to trick bees in the hive – it sounds like the high-pitched noise a queen bee emits when she is instructing her workers to stop moving. That does sound like it would be a useful adaptation for a honey thief.

Returning to mimicry involving distasteful species, there is a different mechanism which has resulted in groups of unpalatable species all mimicking each other. It is called Müllerian mimicry. Unlike Batesian mimics, where there is a distasteful model and a perfectly edible mimic, Müllerian mimics are all distasteful. All the species in a Müllerian system look alike so that it is easier for predators to learn that their warning colours means they should be left alone.

This system helps to explain why stinging bees and wasps are similarly coloured, and also why all the butterflies on the right hand side of this case are so similar. They are four different species of *Heliconius* butterflies, all of which taste disgusting, and that's why they have the classic orange and black warning colours.

In a Batesian system, distasteful models are being mimicked by tasty ones – this is bad news for the models as it makes it less likely that predators will learn not to eat them. Conversely, because distasteful Müllerian mimics converge on a similar set of warning colours, the strength of that warning is increased. This makes it easier for predators to learn the warning signal and avoid eating them.

Having more animals with 'honest' warning colours is a sensible evolutionary solution, even if they belong to different species that simply look alike. If the predator learns to avoid one species with the warning colours, it will have learned to avoid all the species that look like it. This puts evolutionary pressure for all the distasteful species to converge on a similar wing pattern – they are all mimicking each other.

This object is particularly wonderful as it shows both Batesian and Müllerian mimicry at the same time. On the left are four different species of *Melinaea* butterflies – they are from a different genus to the *Heliconius* butterflies on the right (so the four on the left are more closely related to each other than they are to the four on the right, and *vice versa*). *Melinaea* butterflies are all edible, but each of them are Batesian mimics of the inedible specimen to their right, benefiting from the *Heliconius'* warning colours that predators have learned to avoid. Mimicry can be incredibly specific, and each of these species have almost perfectly copied the wing patterns of their distasteful counterpart.

44 Amphisbaenian: A 'Worm-Lizard'

PRESERVED SPECIMEN

Biogeography: Dispersal

The science of where species are found is called biogeography. It incorporates the climatic and ecological factors that determine a species' adaptability to a given environment, as well as the means in which a group's ancestors could have reached their current distribution over geological time.

The distribution of species has been of longstanding interest to biologists. Why is it, for example, that some species live across huge tracts of the globe, while others are restricted to tiny ranges? Why does, and how can, the grey wolf live across the Northern Hemisphere from Western Europe, eastwards through Asia and across North America, while the closely related Ethiopian wolf only lives in a single country? What is it about certain taxonomic groups that means they are found in certain kinds of ecosystem, but not others?

Evolutionary considerations now play a major role in answering these questions, as

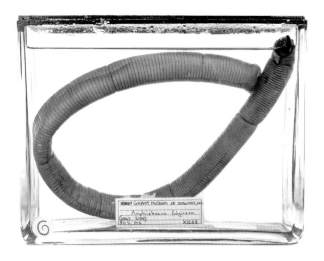

evolution introduces time as a factor in explaining how animals got to where they are. Before evolutionary explanations were understood, the prevailing assumption was that species were divinely created for the places they were found, and may then spread locally from these 'centres of creation' to increase their range slightly into nearby areas.

Darwin made some striking observations that led him to challenge these religious explanations as he voyaged around the globe on HMS *Beagle* between 1831 and 1836. He noted that certain groups of animal appeared absent from remote oceanic islands. Why, he asked, would a creator choose not to furnish these islands with amphibians or terrestrial mammals, despite the fact that there seemed to be ample habitat suited to them? He noted that they could be found on small islands close to land, but not distant ones that were otherwise similar.

Alongside his own observations, Darwin searched through the accounts of previous expeditions and found the same thing: for islands that were more than 300 miles away from a continent or a large continental island, the only records of mammals he could find were for domesticated species linked to humans, or bats. Take New Zealand for example: this is a country with no native land mammals except for three species of bats (though one has most likely recently become extinct). In *On the Origin of Species* (1859) Darwin enquires:

> Why, it may be asked, has the supposed creative force produced bats and no other mammals on remote islands? On my view this question can easily be answered; for no terrestrial mammal can be transported across a wide space of sea, but bats can fly across.*

Darwin was talking about dispersal. Dispersal is now known to be one of the principal forces of biogeography: that species can travel from the place in which they first evolved and spread to new areas. This can be through active travel, like flying, swimming or walking; or as a passenger on the wind or sea currents. Needless to say there is a lot of luck involved with setting off aimlessly from a continental beach, for example, and landing on a distant shore. This explains why amphibians rarely reach remote islands: their skin cannot deal with prolonged exposure to salt water.

Once they have dispersed to their new homes, if they manage to become established, individuals are then reproductively isolated from the rest of the members of their species back on their native soil. This is prime opportunity for speciation to occur, both by natural selection favouring traits that are particularly suited to the new habitat, or by the

---

* Darwin, C., *On the Origin of Species by Means of Natural Selection* (J. Murray: London, 1859)

process of genetic drift. This is a powerful evolutionary mechanism which derives from the chance processes inherent in dispersal events. By statistical chance (as well as some biological factors), the few individuals that arrive in the new locality will not represent all the variety among the original population – some varieties will not be present, and some will be present at levels higher than 'back home'. In this way the new population will 'drift' towards the characteristics of the new pioneers. It's called the founder effect (see chapter 76). Over time the new population can become a distinct species.

Dispersal doesn't only explain how organisms reach islands, but can also be the reason that related species are found on different continents. It doesn't take a great leap of imagination to understand that given time, a group that first appears in France can spread to India or South Africa by dispersal over land (and perhaps go subsequently extinct in the countries in between, leaving the populations isolated). However, dispersal can also be behind distributions that extend over huge oceans.

Amphisbaenians, or worm-lizards, are an unusual group of cigar-shaped burrowing reptiles (showing similar traits to snakes and legless lizards through convergent evolution). Because they burrow, one might assume that their ability to disperse over significant distances is limited. Yet they are found across the Americas, Europe and Africa. Until very recently, another biogeographical process called vicariance was thought to explain their huge range – that continental drift was responsible (see the next chapter).

However, a 2015 study led by Nicholas Longrich at Bath University looked at the fossil record and DNA of this group, and found that the truth is genuinely amazing. A major tool of modern evolutionary biology involves comparing the genetics of different species to calculate the time since the branches of their evolutionary tree split from each other. The earliest fossils are from North America, and the dates Longrich's team obtained from the fossils and genetic evidence confirmed that continental drift could not explain the distribution of worm-lizards across the globe.

Instead, the group must have dispersed across the ocean by floating on rafts of vegetation which break away during massive storm events, carrying anything in the soil or on the associated plants with them. This may sound farfetched, but 'rafting' is actually quite a common mode of dispersal, and given the millions of years involved isn't as unlikely as it may seem. What's brilliant about this story is that relationships between the worm-lizards on each continent show that they crossed the Atlantic at least three times: from North America to Europe, from North America to Africa and from Africa to South America.

45 Essex Emerald Moth

PINNED SPECIMEN

Biogeography: Vicariance – Travelling Without Moving

The counterpart to dispersal (see previous chapter) is vicariance. Together they form the two chief mechanisms by which groups of animals become distributed across space. Vicariance is when a species' geographical range is split apart by a new geophysical boundary. Finding themselves on opposite sides of this barrier, the two new populations are reproductively isolated from each other – the starting point for new species to evolve.

Having been isolated by the creation of the English Channel at the end of the last ice age, the Essex emerald moth developed into a new sub-species.

The barriers that cause vicariance are often the result of huge geological processes operating on an epic scale. Among the most significant of these is continental drift – the action of the tectonic plates which make up the earth's crust slowly but powerfully moving giant chunks of land and seabed across the planet's surface. Depending on whether neighbouring plates are moving away from each other or being forced into one another, continental drift can either build oceans or mountains. Both of these can form barriers that separate once contiguous populations, resulting in vicariance.

Finding both fossils and living species from the same taxonomic groups on two different continents separated by an ocean was a key piece of evidence in proving that those continents were once united. As we have seen, dispersal can also explain how a group can be found in two distinct regions, but by tracing the fossil history of the group in the two localities, palaeontologists can discern whether dispersal or vicariance was at work.

One of the most famous examples of vicariance is the fossils that are shared between the continents of the Southern Hemisphere. The impressive Triassic therapsids are a group which have traditionally been called 'mammal-like reptiles', as they evolved from reptiles and would eventually give rise to mammals. Our ancestors are among these large carnivorous therapsids, which arose in the Permian, 275 million years ago. Cynognathus was a therapsid whose fossils have been found in a band across the middle of South America, as well as central West Africa, dating from around 240 million years ago.

Similarly the therapsid Lystrosaurus (chapter 28) is found in rocks around 250 million years old from Africa, India and Antarctica. How did they end up in such distant continents? These fossils help show that for a period up to around 180 million years ago, all of the southern continents were joined together, in a great supercontinent called Gondwana. It comprised the land that today we call Africa, South America, India, Australasia and Antarctica.

As these landmasses were all conjoined, there should be no reason that organisms living at that time did not occur across more than one of the areas that make up the separate continents we see today. The rifting apart of the huge tectonic plates over geological time has since put oceans between them, but the evidence that they were once united can be seen in the fossils they hold as well as the groups that live there today.

Continental drift is one process that can put seas between lands that were once connected, but climatic change is another agent of vicariance. When global temperatures are low, as they were during the various ice ages of the Pleistocene epoch (around 2.5 million to 12,000 years ago), much of the world's water is captured in the polar ice caps, making the sea levels much lower than they are today. At these times lands that are normally separated by shallow seas are united, allowing species free movement between them. Once the ice melts in the periodic warm phases (we are living in such an 'interglacial period' now), the seas return to flood the channels and separate the fauna living on each landmass.

All across the world there are neighbouring countries that are united by land bridges when sea levels drop, but become isolated when the ice melts. In the cold phases New Guinea connects to Australia, Russia unites with North America, the islands of Southeast Asia all become linked to the continent and Great Britain joins up with Europe.

These cold phase land bridges allow species to disperse and spread into the newly connected territories, however when the sea returns and recreates the barrier the distinct populations become isolated. This is what happened when the English Channel reformed after the end of the last ice age (as it had a number of times during the cycles of glaciations through-out the Pleistocene). The area between East Anglia and the Netherlands was dry until around 6,500 years ago.

The relatively brief (geologically speaking) length of time since then does not seem like it could be long enough for distinct species to evolve on either side of the Channel – indeed Great Britain only has one species of vertebrate it can call endemic (which means it evolved here and lives nowhere else), and that's a bird called the Scottish crossbill. However, sufficient time has passed for a number of sub-species to evolve.

As with other 'vicariance events', the creation of the English Channel divided members of the same species, allowing them to follow different evolutionary trajectories. The designation of sub-species by taxonomists is less common among vertebrate groups (though the Scottish wildcat is considered a separate sub-species to its continental cousins), but there are a number of endemic insect sub-species (most insects have much shorter lifespans than vertebrates, which means evolution can work faster). One of these is the Essex emerald moth, *Thetidia smaragdaria maritima*.

Sub-species are considered taxonomically distinct, but not quite different enough to be called a full species. They are reproductively isolated from other sub-species, and so could one day – given enough time for evolution to work its course – become a full species. The Essex emerald belongs to a species of a moth that is found in a number of European and Asian countries, but its sub-species only lived in south-eastern England. The creation of the English Channel separated it from its continental cousins, allowing it to evolve some unique characteristics.

Sadly the Essex emerald has not been seen since 1991 – it is presumed extinct as a result of land-clearing and overgrazing on its saltmarsh habitats. Maybe one day it would have been isolated for long enough to have formed a completely distinct species, but now we will never know.

46 Nine-Banded Armadillo

SKELETON AND CARAPACE

Biogeography: Geodispersal – Breaking Down Barriers

Just as barriers are built by geophysical processes, they are also broken down. Sea levels drop to reveal land bridges between neighbouring land-masses, and plate tectonics pushes continents together. Geodispersal is the merging of the fauna and flora when these barriers fall, and it is an important process in biogeography – the study of the distribution of species.

The world is constantly changing, as continents drift, oceanic islands form, ice caps fluctuate and sea levels go up and down. Geodispersal, vicariance and dispersal all go hand in hand to determine the distribution of species around the globe. Continents connect and disconnect, and the evolutionary trajectories of the animal groups that are on them at any given time are significantly affected by what happens to the land they live on.

South America and North America have been united and isolated at various points in their histories. The current union begun around 3 million years ago when the Isthmus of Panama was formed. Prior to that, they were last connected over 65 million years ago. This is roughly coincident with the age of the oldest known true marsupial fossils, which were found in North America. From there, they were able to spread down into South America.

Once the two continents separated from each other 65 million years ago, marsupials and their close relatives disappeared from North America, Europe and Asia, but flourished in South America where the ancestors of today's opossums evolved. At this time, the Gondwanan continents of South America and Antarctica were still connected by a land bridge (until about 35 million years ago), and marsupials dispersed south. Antarctica in turn was connected to Australasia, and this is how the famous marsupials of Australia got there, around 55 million years ago. They became isolated there as the route to Antarctica closed with the spreading of the Southern Ocean. This four-continental history is a nice example of how the three biogeographical processes of geodispersal, vicariance and dispersal overlap to contribute to today's biodiversity.

The reunification of North and South America 3 million years ago was a significant geodispersal event in the history of the world, allowing the two faunas to meet after a period of evolutionary separation of over 60 million years. It is known as the Great American Biotic Interchange. (It's worth noting that it is also a significant vicariance event for marine species, as it separated the Pacific Ocean from the Caribbean Sea and Atlantic Ocean. They wouldn't be united again until the Panama Canal was built in 1914).

During that period, North America has been regularly connected to Eurasia (the giant geographical landmass that makes up the politically-defined continents of Europe and Asia) via its north-western route into modern Russia. This meant that many animal groups were shared between the two. Among the mammals, this included cats (with sabre-tooths among them), dogs, otters, bears, horses, deer, camels, peccaries, gomphotheres (odd four-tusked relatives of elephants), and tapirs.

In contrast the animals of South America had evolved in relative isolation, with significantly less connection to the rest of the world. They included the marsupials, the xenarthrans (the group that comprises sloths, anteaters and armadillos) and the meridiungulates: unusual hoofed mammals that are no longer with us.

When the Isthmus of Panama formed 3 million years ago it gave the opportunity for the two communities to mix – the North Americans dispersed south, and the South Americans headed north. Looking at today's fauna, it would be tempting to say that one group was more successful than the other, as there are very many more species of northern origin in South America than *vice versa*. North of Mexico, the Virginia opossum is the only surviving marsupial and the nine-banded armadillo is the only xenarthran, for example. However it isn't quite that simple as initially the migrations in each direction were roughly balanced.

For the few million years between the Great American Biotic Interchange and the end of the last ice age, a little over 10,000 years ago, a number of South Americans were able to establish themselves in the north. Most notably the huge xenarthrans: giant ground sloths and glyptodons – large armour-plated relatives of armadillos that resembled spherical tanks. However, the vast majority of those that headed north were not able to stand the test of time and have disappeared.

The arrival of the North American carnivores like the ancestors of maned wolves, spectacled bears and jaguars in South America did have a big impact on the existing animals there: until then the only large mammalian carnivores had been among the marsupials, all of which went extinct after their North American counterparts arrived. By contrast there were no extinctions in the north directly linked to the arrival of South American dispersers. While the northerners gained dominance among the predators, the southerners were largely able to hold on to their herbivorous niches, including the sloths, but even they decreased in diversity.

On the whole, taking into account the 3 million years since, the groups that originated in North America do appear to have been more successful. One theory for why this is the case is that their long history of interchange with Eurasia made North Americans more competitive than the South Americans, who until then had largely been protected from invasion by their relative isolation.

47 Crab-Eating Macaque

PRESERVED BRAIN

Biogeography: The Wallace Line

Many evolutionary biologists consider Alfred Russel Wallace a bit of a hero. It's not that they don't think that Charles Darwin was legendary, but for many Wallace is a bit hard done by while Darwin gets the lion's share of the credit. The average member of the public has probably not heard of Wallace, and that's why many working in this field will take every opportunity to give him his dues.

MAGACUS CYNOMOLGUS. MACAQUE MONKEY

In the fits of a malarial fever in the Indonesian rainforest, the mechanism by which evolution works came to Alfred Russel Wallace. He, being a humble fellow, wrote to Darwin to seek his thoughts on his theory. This caused Darwin a bit of a shock as he had been working on an almost identical idea for nearly 20 years. We know it as natural selection. To both of their credit, rather than rush to gazump the other by

Macaques are among the few species that were originally from Asia but have managed to spread eastwards across the Wallace Line.

157

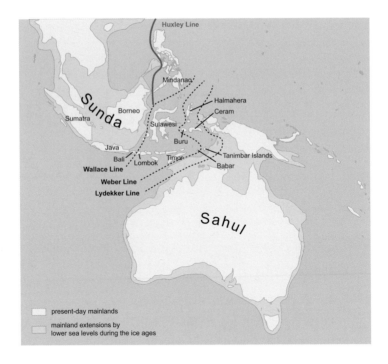

publishing first, in 1858 a paper co-authored by both Wallace and Darwin was presented to the Linnean Society in London. It was arguably one of the most important moments in the history of science.

For a number of reasons it was Darwin who came to be famous for their evolutionary theory, and Wallace has somewhat been left in the shadows (for one, Wallace was still in Indonesia). However there is one area where Wallace has really left a mark: he has been described as 'the father of bio-geography'. Much of what we know about the science of where species are distributed is thanks to Wallace. He has a very interesting faunal boundary named after him: the Wallace Line marks the meeting point of the Australasian and Asian ecological zones.

Unlike Darwin who was a man of independent means, Wallace had to earn a living. He did this chiefly by collecting exotic zoological specimens to be shipped to the UK for sale. Between 1854 and 1862 Wallace was collecting specimens in Indonesia, Singapore and Malaysia. He wasn't just a commercial trader, however: he also used his specimens to research and publish on the area's natural history.

Wallace was struck by how island of Bali shared nearly all of its bird species with Java – the island immediately to its west, but relatively few with Lombok, just 35km to its east. The mammals observe a similar trend, and the same pattern is also seen between the islands of Borneo and Sulawesi further north. If a line is drawn, running roughly northeast to southwest, between Bali and Lombok and Borneo and Sulawesi, Wallace noticed that the species to the west are typically Asian, but to the east they are typically Australasian. Given that these species were distributed well among the islands and mainlands in their respective areas, why had so few of them made the jump across this tiny gap?

During the ice ages, when water is trapped at the polar icecaps, the sea levels are considerably lower. In this region land up to 120m below today's sea level would have been dry land. This united the islands of Southeast Asia with the Asian continent – one could walk from Bangladesh to Bali without getting wet feet, but no further. Similarly the great island continent of Sahul comprised Australia, New Guinea and a number of the chains of islands to its northwest. The remaining islands of 'Wallacea' have not been connected to the mainland, living perpetually in isolation and only reachable by flying, swimming or floating on rafts of vegetation.

The earth's crust which follows the Wallace Line, however, is a significantly deeper chasm. It is the boundary between two continental plates, and runs to a depth of 250m so it is never uncovered by falling sea levels. Without the periodic opportunity to cross a land bridge between islands either side of the Line, Asia's primates, big cats, rhinos and deer dominate to the west of the line. To the east, marsupials and monotremes make up the bulk of the mammals and the two faunas hardly mix. Despite the fact that they can fly, many birds will not deliberately cross open stretches of water, and so a similar trend is seen among them too.

Nature rarely lets a rule get in its way, and of course there are exceptions. Macaques are some of the most widespread and adaptable species of primate on earth, and a number of them have crossed to live to the east of the Wallace Line among the Australasians.

One such species – the long-tailed macaque (also called the crab-eating macaque, which is rather misleading as they don't eat a lot of crabs) – is a champion disperser, having the most discontinuous range of any of the twenty-two macaque species, and the second largest overall distribution. It has colonised many of the islands in the region, but as the fossil record and genetic studies suggest that it likely dispersed over land bridges, how did it get across the Wallace Line?

Humans seem to be the culprits here – it is believed that people introduced crab-eating macaques by boat to islands east of the Wallace Line around 4,000 years ago, potentially confusing today's biogeographers.

There are macaque species that are only found on the Australasian side of the line – suggesting their ancestors must have made it across the deep straights of their own accord, presumably by accidental rafting. The black crested macaque is found only on the north-eastern tip of Sulawesi. Perhaps it's worth noting that this is the first island you reach when crossing the Wallace Line, but it's still a testament to the adaptability of macaques that they have managed to gain a foothold in a new biogeographical zone.

48 *Megaladapis:* The Koala Lemur

Adaptive Radiation: Variations On A Theme

Imagine living in a forest near a river during a storm. A combination of wind and current dislodge a large chunk of the riverbank, as well as all the trees and shrubs, and everything that lives on them. Including you. As the river washes your wretched home downstream, you wish that your peculiar raft would snag on something so you can jump off, but as you reach the coast all hope fades. Eventually you are at sea, and out of sight of land.

Koala lemurs were just one of the species of giant lemur that have recently become extinct.

With no fresh water on your floating island, except for what falls from the sky, you and your fellow castaways don't expect to survive for long.

A month passes, and many are dead, but not everyone. Those of you who are still clinging on cannot believe your eyes when you spot trees on the horizon. You drift closer, and eventually – against all odds – you make landfall on a foreign shore.

You find there is no one like you on your new home – you are a pioneer. You set about making the most of the resources you've found there. Your numbers grow and you begin to spread out, into new regions. You diversify: members of your community start exploiting different ways to make a living, and you start to live different kinds of lives …

… And 50 million years later there are a hundred different ways of being like you; all of them descended from the handful of drifters on your raft.

This is the most likely version of the story of Madagascar's lemurs.

In recent years 100 species of modern lemur have been described. Some of them are controversial as their differences can only be detected genetically – they look anatomically identical – but in any case lemurs are incredibly diverse. They could represent up to 25 per cent of all primate species, and yet they are only native to a single country (albeit a big one): Madagascar.

Continental drift had split Madagascar away from Africa by 120 million years ago. There is some controversy over the age of the lemur group: fossils suggest an origin something in the region of 55 to 35 million years ago, while molecular data predict a date up to 20 million years older than that. Either scenario means that lemurs split from other African primates long after the Mozambique Channel formed, isolating Madagascar in the Indian Ocean. Therefore lemurs could not have colonised over land. The chance drifting of a raft of vegetation is the most likely explanation.

Lemurs are considered by most primatologists to be a solid taxonomic group, with a single common ancestor. This means that there only needed to be a single 'invasion event', like the one described above, to bring about lemurs' current diversity.

When an organism newly appears in an environment and evolves into a number of different forms to fill a range of different roles there, it is called 'adaptive radiation'. Perhaps the most famous example of adaptive radiation is Darwin's finches in the Galápagos Islands. Darwin noticed that on the different islands that he visited, different species of finch were found, occupying a slightly different niche. They were clearly closely related, but were slightly different in the shape of their bills – each adapted to a completely different ecological role largely depending on what they ate. Some had solid seed-cracking bills, while others' bills were petite for picking up insects, for example.

Lemurs are an even more remarkable example of adaptive radiation, having evolved into perhaps 100 different species. In the absence of much competition on Madagascar, and with mammalian predators only arriving around 20 million years ago, lemurs radiated to fill a huge number of empty niches there. They include the smallest of all primates – Madame Berthe's mouse lemur weighs just 30g – and until very recently a species that was in many ways much like a gorilla: the extinct *Archaeoindris* weighed nearly 200kg.

Frustratingly, Madagascar has an awful mammalian fossil record. The only rocks that are younger than the arrival of lemurs are Pleistocene in age: less than 2.5 million years old. However, recent deposits (too recent to have fully fossilised) show us that within the last 2,000 years a wide range of huge lemurs have disappeared from the island. This date coincides with the arrival of humans. Since then, seventeen species of lemur have become extinct, all larger than the ones that have survived. Among them were some incredible examples of convergent evolution.

Aside from the 'gorilla lemur' there were 'sloth lemurs' (the palaeopropithecids), and my favourite, the koala lemur *Megaladapis*. They were like koalas in their ecology and proportions, except they were five times as big. It's incredible to think that such a species existed less than 700 years ago. To put it into some context, they were still around when Cambridge University celebrated its 100th anniversary.

Still with us is a fantastic diversity of lemurs. There are a number of generalists, but many have evolved to fill quite specific roles in the ecosystem.

Some graze, some are nocturnal, some eat insects and some eat seeds. This single group have differentiated to fill the roles occupied by a number of different orders elsewhere in the world.

Sad to say, however, that they are now ranked among the most endangered species on the planet. Despite all descending from that presumably tiny gang of survivors which floated on a raft of storm-churned vegetation, their incredible journey could soon be over.

The extinct koala lemur *Megaladapis*.

49 Hoolock Gibbon

MOUNTED SKELETON

Adaptation

We don't have to look very far into the animal kingdom to find ingenious feats of engineering that form superb adaptations to a species' way of life. It sounds like an odd thing to say, but I find the human arm pretty remarkable. Indeed the freeing up of the ancestral human arms, as we began to walk on two legs, has been proposed to be the key driver in the

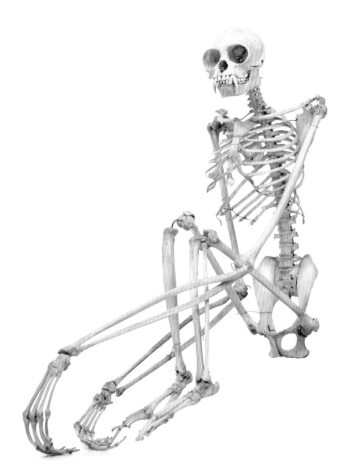

evolution of the human brain: our free hands can create tools, and tool innovation needs a big brain.

Starting at the top, we have a ball-and-socket joint at the shoulder. This allows the entire limb to be moved in an arc motion, creating circles of any size and reach out in any number of directions; it can also rotate, so with your arm out straight in front of you, you can twist your upper arm around 90 degrees. Next, the elbow is a tight hinge joint, allowing a strong, stable movement in a single plane. Combined with the rotation of the shoulder, this allows you to bring your hands into your upper body whilst carrying heavy loads, and adjust the position of your hands for close manipulation.

The two bones of your lower arm are the neatest trick: one end of the lower one (the ulna) is locked into the solid hinge joint of the elbow, while the upper one (the radius) can twist around it. This combination provides both strength and manoeuvrability. Altogether the muscles and joints of the arm allow the hand to be twisted nearly 360 degrees. Finally, the small bones of the wrist and hand, and their muscles, allow incredibly fine manipulations – so precise that we can thread a knot in silk, write on a grain of rice or play a piano.

That said, with the exception of a relatively small number of things (if you think about it), anything we can do, animals can do better. Arguably the most remarkable arm or forelimb in the animal kingdom belongs to the gibbons.

Gibbons are a group of around nineteen species of small apes, all of which are found in Asia. Humans, being a hierarchical species, occasionally refer to the gibbon family as 'lesser apes', which doesn't really mean much biologically and is simply insulting. Such denigration risks making people think that gibbons aren't very important, which is a very dangerous mindset as they are the primates most in need of our help. Every single species is threatened with extinction, and the Hainan gibbon may be the rarest wild mammal in the world. Less than thirty of them are eking out an existence in a tiny pocket of rainforest in China.

Gibbons are perfectly adapted to life in the trees – they can move through the high canopy faster than any non-flying animal – over 50km per hour. If you consider the complex habitat – the irregularly angled branches, huge distances involved and many obstacles, let alone the fact that they are tens of metres in the air, this is pretty astonishing. They do this with a unique mode of locomotion called brachiation, which is swinging hand-over-hand from branch to branch without the use of legs or a tail (and being apes, they have no tail). The adaptations of the gibbon arm allows them to swing from thin branches, far from the tree trunk, where they can reach more of the fruit they like to eat.

The overall layout is similar to our arms, but the most obvious difference is that gibbon arms are almost comically long. On the occasions that

A white-handed gibbon brachiating with a baby.

they do venture down to the ground they have to walk with their hands in the air to keep them out of the way. Increasing the length of each bone (and particularly the forearms) allows them to attain a giant reach. This is essential for moving between the sparse branches of the canopy. Although their bodies are extremely lightweight, reaching far forward with one arm creates an extremely efficient pendulum system – swinging their bodies past their hand grip and on to the next one. Despite the speeds they can reach, brachiation requires surprisingly low amounts of energy.

The pendulum effect is increased by concentrating most of the bulk of the arm muscles as near the body as possible (pendulums swing best when the mass is all at one end). It also moves the weight away from the hands, meaning the non-supporting arm can move more quickly and with more power into the next position, as it has less to lift. This arrangement requires them to have long tendons running down their arms, which are lighter than muscles and create high forces and store elastic energy, which can be used in the next swing cycle.

They have very strong flexor muscles at the elbow and wrist (those which make the joints bend), which can change the arc of the pendulum, and carry the weight of the animal on a bent arm. It's worth noting that being suspended down from your arms means that gravity works on gibbon joints in the opposite way to animals that have to hold themselves up from their legs – it has taken quite a bit of evolutionary alteration.

On the skeleton, one of the most significant adaptations is that their wrist joints function like a ball and socket, allowing them to grab any branch they approach, whatever direction it is pointing without changing the direction of the swing. Their fingers are also extremely long and per-manently curved, to increase grip. A tight hold is very important at gibbon height, and they have fused some of their muscles there, which increases their power. This has given them less manual dexterity, so they can't move all of their fingers independently. That's a small price to pay for having the freedom of the treetops.

50 North American Beaver

SKULL

Adaptation And Skull-Reading

At school, many children learn that it is possible to look at an unknown skull and work out something about how the animal lived. They know that a collection of sharp stabbing canines and long blade-like molars suggest the owner is a carnivore; that flat, cutting incisors and bumpy, grinding molars belong to herbivores; and that a mix of the two is found in omnivores. If their teacher is particularly conscientious, pupils will learn that predators have the eyes on front of their head to help judge pouncing distance, while prey species need a wide, watchful field of view, so their eyes are on the side.

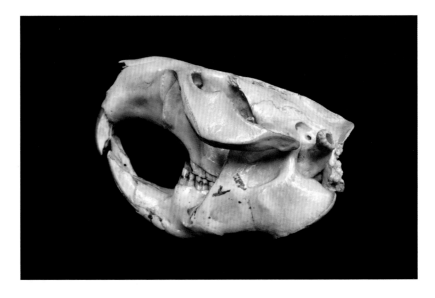

These are really simple ways to 'read' a skull, but there are plenty of other clues to look for to learn about an animals' lifestyle, before their identity has even been disclosed. The teeth, lumps, bumps, smooth patches and grooves can all be critical adaptations to allow the animal to make the most of its environment. They have been honed by evolution for a specific purpose. The beaver has one of my favourite skulls, packed full of evolutionarily significant traits to help the beaver be a beaver.

Beavers are rodents so they have massive, ever-growing, self-sharpening front teeth. Rodent incisors are often differently coloured on the front and back. The orange substance on the front side is extremely hard enamel, while the back is dentine, which is a softer material that is normally found on the *inside* of more 'normal' teeth (including ours) – it is unusual to find it exposed like this. When rodents bite down on hard material, or even by biting their top teeth against their bottom teeth, the dentine at the back of the teeth erodes away at a faster rate than the enamel at the front, essentially sharping the 'blade' at the tip.

Given that they are constantly grinding away bits of their teeth to keep them sharp, they have an amazing adaptation that allows them to regrow continuously. Their teeth are open-rooted, unlike ours which have roots that come to a definite end point. For us, the cells which deposit enamel (ameloblasts) die once our teeth have developed, but for beavers they keep beavering away throughout the animal's life. This allows for the constant wear that results from biting hard wood, by continually replacing the enamel from the bottom.

Beavers are famous for chopping down trees, and they do it incredibly well. I have watched Swedish beavers topple trees by gnawing through trunks that that are considerably wider than my shoulders. While beavers chew through a lot of wood, they do not eat it. However there are a number of reasons why they cut down trees.

Most dramatically, beavers fell and drag trees to use the timber to build dams across streams and rivers. This creates large wetlands and even sizeable lakes. Beavers are known as 'ecosystem engineers' because they are essentially creating their own habitat. The water allows them to safely move around to access their food. The ponds act as moats around their lodges where they shelter (which are also made of felled trees), protecting them from predators. Beavers eat the bark, shoots and leaves of the trees they cut down, and also the water plants that grow in the wetlands they create.

A skull that has huge gaps between the cheekbones (zygomatic arches) and the skull proper belongs to an animal that needs a lot of space for the sizable jaw muscles. In the case of the beaver, cutting through wood obviously requires a strong bite.

Muscles need a solid point on a bone to attach, and that often takes the form of a ridge on the bone. The bigger the ridge, the bigger the

muscle. The large ridge of bone at the top (the sagittal crest) of beaver skulls increases the surface for their powerful temporal muscles to attach. Also, the cheekbones themselves are really deep and wide, allowing for extra huge masseter muscles involved in chewing. This gives them a very distinctive looking zygomatic arch, really unlike any other.

When you swallow, a flap of tissue at the back of your throat (the epiglottis) flips over the top of your windpipe (trachea) to stop things getting into your lungs. Unusually, a beaver's epiglottis is at the back of its nose, not its mouth. A beaver can hold the back of its tongue tightly against its palate, blocking the passage of water from the mouth. This allows the beaver to open its mouth underwater, to gnaw or carry branches – a crucial trait for being a beaver. This isn't visible on the skull, but it's still pretty neat.

If you ever see a skull with its eyes and nose on the top of their skull, it's a good sign they spend a lot of time in the water, but need to function out of it. Beavers swim with the tops of their heads just above the water, so they can see and breathe and hear. I noticed that the bony tube of the ear (the external auditory meatus) on a beaver skull extends upwards and outwards in a way that is far more pronounced than in any other skull I can think of. It gives them the appearance of alien-like antennae.

In 'reading' this skull, I assumed that this was also an adaptation to the aquatic lifestyle; raising their ears above the water. However, my colleague pointed out that this feature isn't shared with coypu, capybara or water voles, which are all aquatic rodents, but the extinct beaver *Palaeocastor* does share the pronounced ear tubes although it didn't swim. He went on to hypothesise that the beaver's meatus wasn't adaptive at all, it was merely a means of allowing space for its big jaw to move without blocking the ears, as rodents chew backwards and forwards, not up and down. This just goes to show that it's very easy to make assumptions about what anatomical features are for (zoologists call an unverified adaptational theory 'Just So Stories', in the style of Kipling's *How the Leopard Got His Spots*, for example). As much as reading skulls is a useful skill, it's important to look at the fossil record and the living relatives before coming to too many conclusions about why an animal is built the way it is.

51 Hippopotamus

Sexual Selection: Big Teeth

This is another animal whose teeth aren't just used for eating. In fact, hippopotamuses and beaver skulls share a lot of similar features, but only some could be considered examples of convergent evolution. Like the beaver (see previous chapter), hippos' eyes are on the tops of their heads. They actually sit in little turrets that pop out above the top of the skull. Likewise their nose and ears are high on the skull like a beaver (and a crocodile) – these are all convergently evolved adaptations to spending lots of time in water. By contrast there are other features, for example massive front teeth, which may appear similar but have evolved for very different reasons. Again, like the Just So Stories, this shows the potential difficulties of inferring ecology from quick assumptions.

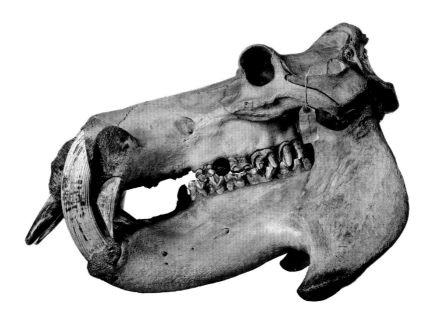

Hippos have the longest canines of any land vertebrate. There are large solid 'tusks' on both the upper and lower jaws, but the lower ones are much bigger. In males they can be up to 30cm long, with another 40cm hiding below the gum (females' are slightly smaller). Like a beaver's incisors, they are self-sharpening, but they don't have the shrewd rodent adaptation of a softer dentine backing to make the sharpening easier.

The upper and lower canines grow into each other. As hippos bite down the solid enamel of one shears against the surface of the other, so they are both flat where they meet. This constantly removes a layer of the teeth and keeps them sharp. Like the beaver, this means that the teeth need to be open rooted, growing constantly throughout their lives. If a hippo loses a canine, without its partner to shear against, the tooth in the opposite jaw will continue growing in a massive spiral curving into the face, which can eventually seriously inhibit their ability to feed.

Their incisors are also long – up to 17cm long (humans' are less than 1cm) – with open roots. They have two pairs of these long peg like teeth in each jaw (though pygmy hippos only have one pair in the lower jaw), sticking forwards out of the face.

Large canines are well known to be the classic mark of a carnivore, and hippos have the largest of all. However, hippos are strict grazers – they eat grass (I should say that there are some very rare accounts of hippos scavenging from carcasses, presumed to be related to dietary stress). Their front teeth are clearly not very useful for cropping grass, which is what other grazers like cattle, rabbits and kangaroos do with their incisors. Instead, hippopotamuses use their thick mobile lips to pluck the grass for feeding.

A quick note on nomenclature: there are two living species of hippo – the enigmatic pygmy hippo of Liberia and some surrounding countries, and what is generally called the common hippo. However there is a school of thought that names with the word 'common' in them (common hippo, common wombat, common dormouse etc.) can be counterproductive for species that aren't actually that common. There seems to be a movement to refer to them as the 'large hippopotamus' instead. I should also say that while 'hippopotami' is widely used as the plural, it is not technically correct as the name comes from two Greek words (translated as 'river horse') and Greek words ending in '–us' pluralise to '–odes'. It is Latin words that switch to '–i'. So if people are going to be pretentious about it, the plural of hippopotamus is hippopotamodes, but that's just silly. Let's stick with hippopotamuses. The same is also true of platypuses and octopuses, which also have Greek roots.

Back to the teeth. Neither the canines nor the incisors are used in feeding (though it is thought that the incisors might assist in the digging for mineral-rich soils at salt licks). Instead, they are famously used for

territorial fighting. Bull males control a stretch of river – and the females in it – which requires them to defend their territory from potential rivals. The females will also use them to protect their young against predators. The teeth essentially function like antlers or horns as sexually selected weapons. There are actually a number of strictly herbivorous mammals that have significant canines (particularly among primates), so it's important to remember that not everything with big 'fangs' eats meat. It's particularly telling if the males' and females' canines are differently sized.

Before they come to blows, hippos will perform incredible 'yawn' displays to try and convince each other that they are too big to pick a fight against. The jaw joint and musculature allows them to open their mouths to nearly 180 degrees, which is remarkable. At rest, the *orbicularis oris* muscle that curves around the corners of the mouth is highly folded to stop the cheek from splitting when they yawn like this (that's the bit of your cheek that hurts when you stretch it by opening your mouth too wide). The skull also has a pronounced sagittal crest – again like the beaver – to allow for large temporal jaw muscles, which controls the gape.

If the open-mouthed displays don't frighten off their challenger, however, they will do battle. The two hippos clash mid yawn. It is thought that the outward-facing incisors are used to try and hold their opponent's mouth open, while the giant canines are used for stabbing. In this respect, they are more like a deer than a beaver.

52 Alligator

Challenging The Reptilian Stereotype

I think it's fair to say that most people would consider alligators, crocodiles and gharials (together called crocodilians) to be the typical reptile
– they fit our ingrained stereotype for what we consider to be reptilian.
If I asked you to picture a standard reptile, I expect the distinctive features that are generally held to characterise the reptilian way of life would
include a cold-blooded animal which dragged its belly and tail along the
ground; generally sluggish, but capable of short bursts of speed.

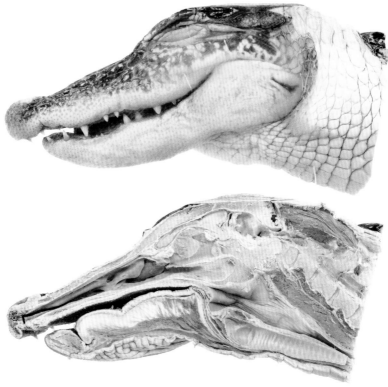

Crocodilians would fit the bill, along with pythons, turtles and at least the historical cartoon versions of dinosaurs (the image of the dozy, tail-dragging, sluggish, cold-blooded dinosaurs is slowly being replaced by an intelligent, active, probably warm-blooded animal with its neck and tail held out horizontally to the ground, and in many cases covered in feathers). However, although crocodilians could be argued to fit the stereotypical view of a reptile, by looking at their evolutionary history we might see a different picture.

Crocodilians originated from small terrestrial sauropsid reptiles typified by the Triassic carnivore *Gracilisuchus*, which walked on two legs and was only around 30cm long. When we picture reptiles today, the standard form is to have what is called a sprawling gait: their limbs are held sideways out from the body, bent at the knees and elbows. This arrangement puts a huge amount of pressure on these joints, as they have to hold the weight of the body at 90 degrees, and gravity works against the joints and muscles to lift the body off the ground. This is basically what happens to your arms if you do push ups.

The first members of the group from which crocodilians, pterosaurs and dinosaurs (and therefore birds) evolved – the archosaurs – did not possess this arrangement. Instead they had an erect gait, like us mammals. Here the limbs are held vertically below the body, with straight limbs. It is argued that this system, which is kinder on the joints against the strains of gravity as the forces pass straight down their length, allowed the dinosaurs to grow to such gigantic proportions.

At first appearance it seems that through their evolution crocodilians secondarily lost the erect gait, and returned to a system with bent knees and elbows. However, we shouldn't write them off as belly-dragging and sluggish. Modern crocodiles are actually capable of four different styles of moving on land: the belly run, where they quickly wriggle down riverbanks by pushing off with their back legs; the 'reptilian' sprawling gait; the high walk, where they can tuck their limbs below the body and run fast with an erect gait much more like a mammal; and perhaps most terrifyingly, in a gallop. They can bound like a horse shifting from back to front legs in pairs. So not so reptilian after all, then.

All twenty-three species of living crocodilian are semiaquatic with short limbs, but their fossil record includes many species that were completely terrestrial with long limbs and some even had hoof-like feet. They were believed to have been fast running, capable of chasing down mammals on land. Conversely, there was also a group of croc-relatives – the metriorhynchids – that were completely adapted to living in the sea as fast swimming predators with paddles for limbs, and a tail like a shark or an ichthyosaur.

There is some debate over whether crocs have always been cold-blooded. Some evidence for a warm-blooded ancestry is provided by studying their hearts. Crocodiles are unusual among the reptiles in that they have a four-chambered heart, just like us. This allows for separation of the blood into two circuits: one that pumps with low pressure from the heart to the lungs and back again, to collect oxygenated blood, and one that then pumps this oxygenated blood at high pressure around the rest of the body; returning deoxygenated blood to the heart to start the cycle to the lungs.

In mammals this system is important in maintaining our oxygen-hungry warm-blooded metabolism. However modern crocodiles (which are cold-blooded) have a controllable shunt between the two vessels leaving the heart, which allows the blood in the two systems to mix. This could be a system of allowing them to remain submerged for longer without needing to breathe; constantly recirculating the blood. However recent studies have shown that this shunt develops late in embryological development, suggesting that it wasn't part of the 'original' crocodilian system, providing an argument that they were once warm-blooded.

When most reptiles breathe in through their noses, the air enters the mouth at the front, just below the nostrils. In us mammals, by contrast, it enters at the back of the throat. We have a secondary palate – a layer of bone that separates the air entering the nostrils from our mouths down a little tube. We need to breathe constantly in order to maintain our warm-blooded metabolisms, so our secondary palate allows us to keep breathing even with our mouths full.

Through the evolutionary history of crocodiles we see the hole where air enters the mouth move further back down the palate through time. Modern crocs have a full secondary palate, with air entering the mouth just above the throat. However, it would be hasty to suggest this was evidence of a warm-blooded ancestry. Their secondary palate acts like a snorkel, allowing them to breathe with their mouths open underwater,

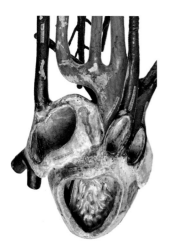

with the tips of their noses just poking out about the surface. It also provides additional torsional strength to the snout to allow them to do the crocodilian 'death spiral', when they rip pieces off their prey by gripping it with their teeth and spinning their whole bodies in the water.

Combining these anatomical features with complex behaviours seen in crocodilians – like elaborate parental care, social hierarchies and cooperative feeding – challenges the stereotypical view of reptiles. Perhaps reptiles are not as reptilian as we thought.

A papier mâché model of a crocodile heart made by Maison Auzoux in the mid-1800s. Crocodilian hearts have four chambers – unlike other reptiles, but like mammals.

53 Gaboon Viper

REPLICA SKULL

Flexible Skulls: Extreme Engineering

Of all the skulls evolution has ever produced, this is my very favourite. It is a genuine marvel in engineering. Snake skulls, and particularly vipers, have reduced the elements of the skull to such an extent that it is basically a strong central tube with a series of peripheral bars that operate as a complex series of flexible hinges and levers.

Human skulls are formed of several different plates of bone joining each other to create a single dome-like unit. In newborn babies, the bones are still growing towards each other, leaving a 'soft spot' between them near the top of the dome. Without moving your neck, try to move as many of the parts of your skull in as many ways as you can.

You should find that, as well as having a tiny gape (we can't open our mouths very wide), human skulls are incredibly immobile: the only part that has any movement at all is the jaw joint, and even then only very slightly. We can put the front of our top and bottom jaws out of alignment by maybe 5mm to the left or right, and create an overbite or underbite of even smaller than that.

Some mammals fare slightly better, particularly herbivores that have to chew the cud. The hinge where their lower jaw meets the skull is very open, allowing the lower jaw to swing from side to side a lot more, facilitating the intense grinding necessary for breaking down tough plant material.

Some mammals also have a bit of flexibility at the front of the lower jaw too. The mandibular symphysis is the point where the mandibles forming each side of the lower jaw meet (in mammals the mandible is only formed of one bone; in reptiles it is seven). In primates – including humans – the mandibular symphysis is very strong – the two sides of the lower jaw are rigidly attached to one another at the chin, giving us a strong, stable bite. In groups which might experience differing stresses on either side of the jaw – for example from tackling large struggling prey – the two mandibles are more loosely connected, which prevents them breaking under stress.

Still, mammal skulls are extremely immobile. By contrast, many animals can move different parts of their skulls relative to each other – a feature called cranial kinesis. In fact, cranial kinesis is reported in all tetrapod classes except mammals.

The true masters of this are the snakes. They have drastically reduced the amount of bone in their skulls by losing some bones and reducing others to simple rods. Unlike in our skulls, where the separate plates rigidly fuse together with interlocking sutures, many elements of snake skulls have joints, hinges or loose connections between them.

Having kinetic skulls has a number of different advantages. As mentioned, it's a lot safer for tackling prey that is still alive and kicking when they bite into it (rather than using hands or feet to subdue prey). Moving parts of the jaw somewhat independently allows for specific movement of the teeth so that they can always be placed tip-first straight into the prey, protecting them from the damage that could be caused by pushing long, thin teeth sideways through food. It also allows teeth to be taken out carefully, providing precise manipulation of food within the mouth.

The downside is that flexible skulls won't allow for strong bites on tough or hard food. Indeed in lizards – which also have a good range of skull mobility – cranial kinesis is lost in herbivorous species so they can chew fibrous plant material.

It is often said that snakes can 'dislocate their jaws'. This is not accurate – instead, they are simply highly kinetic. Mammal jaws only have one point of rotation (on each side): the jaw joint. Snakes, by comparison, have eight. There are eight places that parts of the skull can move relative to their neighbours. That doesn't include their mandibular symphysis, which is completely distensible – the two sides of the lower jaw are only joined by flexible connective tissue.

Each side of the snake upper jaw can be moved outwards. In fact, the bones that bear the teeth are only connected to the solid part of the skull (the braincase) by ligaments and muscles. As the mouth is opened, the teeth on the upper jaws are not only raised and moved forward, but also rotated sideways so that the teeth are freed from the prey, allowing them to move ahead of the next bite.

All of this together means that snakes can make their mouths bigger than their heads – they can swallow animals that are many times bigger than their mouths would be at rest. This is achieved by swinging parts of their skull and jaws outwards around the prey, leaving their eyes and braincase sat on top.

Snakes can move either side of their skulls separately from each other. This is a critical adaptation for cramming large portions of food into their mouths without hands. It allows them to 'walk' their skulls over their food, dragging it into their mouths like a tug-of-war team pulling a rope, hand over hand.

The two parallel rods of bones on the roof of a snake's mouth each have a row of backwardly curved teeth. These teeth hook into the prey and prevent it from getting out of the mouth. By moving one of the rods forwards and unhooking and re-hooking the associated teeth, that side of the head effectively take a step forwards along the prey. Moving the right and left rods in turn results walking the prey down into the throat.

Vipers go even further, and none more so than the gaboon viper pictured here, which grows the longest fangs of all snakes and produces the highest quantity of venom (delivered like a hypodermic needle through a channel running through the fang). When its mouth is closed, the fangs lay flat pointing backwards within the mouth. However, the motion of opening the mouth causes a series of bone levers to pull the root of the fang back, lifting the curved dagger vertically, and ready to strike. Only by adopting this lever system could snakes evolve longer fangs, otherwise they would not be able to close their mouths.

54 Narwhal

TUSK

Why The Long Face? A Dental Mystery

Narwhals are a kind of whale that live in the Arctic Ocean, related to belugas. They spend a lot of time around ice which is the likely reason that they have lost their dorsal fins, as they could cause problems when swimming along the underside of ice sheets. Narwhal bodies are 3 or 4m long, but you may need to add another 2.6m to that to get their total length as the males have these extraordinary tusks pointing horizontally out of their faces. The tusks are highly modified upper-left canines which are straight but grow with a tight left-hand helical spiral. Their tusks grow directly through their upper lip, which sounds odd but I suppose means that they can shut their mouths properly despite having a giant rod sticking out of the front of it.

Narwhals do have an upper-right canine too, but it's generally left embedded in the jaw. However, it's not particularly uncommon to find a male with two full tusks (there is such a skeleton hanging from the ceiling (the traditional museum habitat of whales) of the University Museum of Zoology, Cambridge). In these cases the left tusk tends to be slightly longer, but remarkably they both have a left hand spiral – this means they are not symmetrically developed: generally body parts that grow on opposite sides of the midline are mirror images of each other, and so we would expect the right tusk to spiral in the opposite direction, but it doesn't. Females occasionally grow a small tusk too (otherwise both canines stay embedded in the jaws), but mostly it's a male thing.

What is the tusk for? That's the million-dollar narwhal question. Generally, when there is massive sexual dimorphism in a species (when the males and females are drastically different), the explanation has something to do with sexual selection. This is the evolutionary outcome when one of the sexes has to compete in some way for the ability to reproduce. Male deer have antlers; male antelope have horns; male seahorses have pouches; male gorillas are giant and male peacocks have crazy tails all because of sexual selection. Therefore it's not surprising that historically the most common explanation for the male narwhal's tusk has been for male-on-male fighting over females. The regularly repeated story goes that they would stick their tusks out in the air and use them to fence.

That would be awesome, but there are very few accounts of narwhals using their tusks aggressively. It's possible that they are used in signalling for mates; displaying their handsome face-spike like the peacock's tail rather than coming to blows. To me that's not entirely convincing.

Over the centuries numerous alternative explanations have been offered for these incredible dental spikes. Could they be for ... spearing fish to eat (but then how would they get them off the end of the spear)? To ward off predators? For digging up food? As ice-anchors (walruses use their tusks to hold onto the edge of an ice floe while they sleep, among other things)? To transmit or receive sound (like other toothed whales, narwhals communicate by clicking)? The problem with all of those suggestions is that there would need to be a good reason why tusk-less females didn't need to do the same thing.

Is the answer that narwhal tusks are sensory organs? This was the conclusion of a recent study led by Martin Nweeia of Harvard School of Dental Medicine (who aside from doing major studies involving temporarily catching wild Arctic narwhals is also a practising human dentist). If your teeth react to cold drinks and ice cream, you will know that they can be sensory organs. Below the hard enamel and cement-like dentine is the soft pulp cavity, full of nerves and blood vessels. For us, painful sensitivities arise when teeth are damaged and expose the nerves to outside environments.

Nweeia's team noted that narwhal tusks were covered in little tubules that enabled sea water to penetrate the teeth and directly contact the nerve

cells in the pulp cavity of the tooth (a feature not seen in belugas, their closest relatives). They were able to demonstrate that the tusks could detect changes in salinity in the water around the tusks, giving strong evidence that they are a sensory organ. They also showed that the same genes and arrangement of the nerves were involved in the tusks as in other sensory processing.

Measuring salt levels could be handy for Arctic whales, as an increase in salinity means that the sea is freezing (the freshwater gets trapped in the ice, but not the salt). This puts narwhals at risk of drowning, as they have to access the air in order to breathe. If they can detect salt, perhaps they can also detect other water-borne chemicals too. The researchers proposed that the reason for the sexual dimorphism could be down to sexual selection after all: the males may use their tusks to locate females who are foraging or ready to breed by detecting their hormones or other chemical signals; or perhaps they helped find sufficient food for the calves to eat. The theory isn't without its doubters, and of course these bizarre tusks may have evolved for more than one reason, but it would be nice if the truth of the tooth had been settled.

55 Bushmaster

PRESERVED HEAD

Snakes That 'See' Heat

A 'sixth sense' is something that people say to imply that someone has perceptive abilities beyond the conventional five human senses: touch, taste, vision, hearing, smell. In reality, humans have many more than five senses. To name a few: we can detect pain (nociception); we are aware of the relative position of our movable body parts to each other (proprioception); we have a sense of balance (equilibrioception); we can sense vibrations (mechanoreception); our bodies can detect different levels of various specific chemicals (chemoreception); and we have the ability to detect heat (thermoreception). We all have a sixth sense, and then some.

In nature, some animals can detect electricity; some can measure the pull of the earth's magnetic field and use it for navigation; many can compute minor localised changes in water pressure (fishes' lateral line systems, for example), and some have extremely heightened versions of the senses humans have above.

Some snakes can detect heat to such a degree that they can 'see' the world in temperatures. The English language doesn't really have a useful word for describing what that might be like (for obvious reasons it's rather difficult to describe the neurological outcome of perceiving a sense that we don't have). So we have to settle with putting 'see' in inverted commas.

We can obviously sense heat, but the heat detection mechanism of snakes is not the same as in humans. We could be described as 'feeling' heat, while snakes can 'see' it. Between the nostrils and eyes of pythons, boas and pit vipers there are holes called pit organs. These features can detect the infrared radiation being given off by other animals as heat, providing a thermal image much as an infrared camera does.

A 2010 study led by Elena Gracheva and Nicholas Ingolia (which includes the wonderfully pragmatic line 'Snakes, particularly pit vipers, are inconvenient subjects for physiological and behavioural studies.') found that snake pit organs have tissues that can sense highly localised differences in temperature from objects that are a metre away or less. As infrared radiation is a similar form of energy to visible light, in theory the snakes could be detecting the photons being given from warm bodies (in the way that eyes see light), rather than the heat itself, but this study found that it was the latter.

The organs contain cells that correspond to wasabi taste receptors in mammals. These are the sensory cells that respond to the inflammatory reaction caused by irritating chemicals in plants like horseradish and mustard. In these snakes, the tissues are heated by the infrared energy radiating off their warm-blooded prey or predators. As long as the animal they are hunting or avoiding is a different temperature to the surrounding environment, they will be able to form an image of it even in complete darkness. This is obviously an extremely effective adaptation for predators that do their hunting at night or by following prey down burrows or amongst rocks, for example.

It's interesting that this ability is found in pythons, boas and pit vipers but not in other snakes, as these families are not believed to be very closely related to each other. Pythons and boas – both of which tackle their prey by constriction and are not venomous – are ancient groups of snakes. Pit vipers, on the other hand, only appeared relatively recently. This group contains rattlesnakes, bushmasters (pictured on p. 181) and moccasins – all highly venomous predators, with significantly more sensitive heat receptors than their constricting counterparts. Although the mechanism used in these snakes is very similar, this suggests that the 'super-sense' has evolved convergently and independently in the pit vipers and boas and pythons, by modifying another sensory system inherited from their ancestors.

Aside from snakes, the only other vertebrates capable of detecting infrared radiation are vampire bats. They use sensory areas on the frilly 'leaves' of skin that grow out of their noses to detect hot-spots on their warm-blooded prey. In this way that can locate areas of high blood flow to assess where best to bite to get a good meal of blood. Their heat receptors and associated parts of the brain are not believed to form a full image of relative temperatures like an infrared camera or pit vipers, boas and pythons. They can, however, determine the warmest part of their host.

56 Electric Ray

PRESERVED SPECIMEN DISSECTED
TO SHOW ELECTRIC ORGAN

Making Sparks Underwater: Electric Fish

So far in this book I have already mentioned some animals that can detect electricity: platypuses, sharks and rays all hunt by sensing the tiny electric charges generated by their prey. All biological cells produce minute levels of electricity. A voltage is just the difference in electric charges between two points (called electric potential). Cells generate an electric potential because their membranes separate different numbers of charged potas-

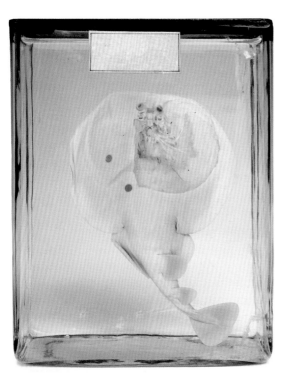

sium and sodium ions inside and outside of the cell. The generation and alteration of these charges is how muscles and nerves fire, for example.

In electroreceptive animals – those that can detect these minute voltages – they have specially adapted sensory cells to locate the source of the electricity: their prey. However there are some animals that can actively *generate* a stronger electric field – electrogenerative species – and they have found a myriad of uses for their electric powers.

Most famously, electric eels use a significant electric field to stun or even kill their prey. This tactic is also seen in the less well-known electric rays (not to be confused with the

An electric eel.

weapons that featured regularly in 1970s sci-fi films). Salt water is an extremely effective electrical conductor, and electric rays can produce a voltage of around 50 Volts, but with a massive current, producing a hefty electrical power of about 1 kilowatt at its peak.

Electrogenerative fish have electric organs to produce their electric discharges. These are typically highly modified muscles that have lost the ability to contract but utilise their electric potentials. They are made of stacks of cells called electrocytes, each of which functions like a tiny battery. Each electrocyte produces a small charge, but they combine in their stacks to add together to produce charges large enough to use. Depending on the conductivity of the water, the stacks then add together in ways which are thoroughly comparable with man-made electric circuits: either in parallel or in series.

In electric rays, which are marine and therefore can benefit from the high conductivity of salt water, the electric organ is made of relatively short stacks (about 1,000 electrocytes in each stack), but these stacks are combined together to fire in parallel. Their electric organs make up a significant portion of their flat pectoral fins, near their heads, and are well connected to the brain. Each cell builds up its electric potential by creating a difference in charged particles either side of its membrane. When the ray detects that prey is in range, it fires all of the cells at once by allowing the particles to cross suddenly back across the membranes, combining to create a massive electrical discharge.

Because of the orientation of the ray's electric organ – lying flat over its fins – the stunning pulse discharges vertically, which matches the way that the ray hunts from its position lying flat on the sea bed. In most other electric fishes, which hunt in the open water, their electrical organs are arranged horizontally along their bodies and so fire horizontally.

Because electric eels live in freshwater, which is less conductive, they have the opposite set-up to electric rays. They have very long stacks of electrocytes – around 6,000 in each stack – which are then bundled together in much smaller numbers. This means that they can fire in series, generating a much lower current but a high voltage of around 500 Volts.

Aside from the species that can generate these sudden large pulses to stun their prey, there is a much wider range of fish that produce a more constant electrical discharge which is used to navigate (these are called 'weakly electric fish'). This includes a number of species of South American knifefish (which, incidentally, is the group to which electric eels belong – they're not really eels), and elephant fish.

These species can detect and identify different kinds of objects by generating a weak electric field and assessing how different items in the environment modify that field as they swim past. Different materials conduct electricity to different degrees, and therefore will affect the fishes' electric fields in different ways. A rock will disturb their electric field in a different way to a submerged plant, for example. This is a particularly useful adaptation for living in very murky environments where they cannot rely on vision to make their way through the muddy waters of the Amazon, for example.

Some electric fishes can even use their electrogenerative powers to communicate with one another. By modifying the charge they give off, they can produce signals used for attracting mates or controlling their territories from passing challengers. Again, these would be useful tools for species that live in murky rivers and lakes.

57 Cuttlefish

Communicating Cephalopods

There is no denying that humans have introduced a hierarchy over the way we generally consider the animal kingdom: some creatures are considered to be lowlier than others. Like it or not, mammals are generally (and perhaps subconsciously) afforded more empathy than any group of animals, and it seems to be human nature for vertebrates to be largely prioritised over invertebrates.

These sentiments are borne out in the way animals are treated in a wide range of situations. Few people would think much of seeing someone squash a mosquito that flew past, but the vast majority would be horrified to observe someone stamp on the head of a rat. It's not difficult to guess why that is. Aside from the size of a rat meaning that its obliteration would be significantly messier, the brutal squashing of a mammal seems cruel because we infer a level of complexity on the rat's 'feelings'. Both directly – we are satisfied that rats can feel pain and experience suffering – and on the impact on other rats: murdering the rodent could result in dependent young starving to death, and most of us would find that idea unpalatable.

Such thinking (and more scientific considerations) is reflected in the laws that protect animals. For example, the UK's Animals (Scientific Procedures) Act 1986 governs how and when animals can be used in research involving experiments that could potentially cause pain, suffering, distress or lasting harm.

Following a couple of amendments, it is specific about which animals enjoy protection under the Act. Mosquitoes are not covered. Rats certainly are.

Specifically, the Act protects 'all living vertebrates, other than man, and any living cephalopod.' Presumably man (and woman) have other legal protections, but it is notable that the only invertebrates to be covered by this law are cephalopods – the group of molluscs that includes squids, octopuses and cuttlefishes. In this way the Act recognises that cephalopods are special – they have complex cognitive abilities including problem-solving and communication.

In ecological terms, communication in animals is normally defined as occurring when one individual uses specially designed signals or displays to modify the behaviour of others. The use of the phrase 'specially designed' is important as it excludes behavioural outcomes that merely result from an observation of information that the first animal did not intend to provide. For example a female frog that is laden with eggs is not communicating to the heron that she would be a good nutritious meal. The frog's visible fatness is not a specially designed signal, although it is likely to modify the heron's behaviour into choosing to eat her.

Cuttlefishes have been found to be capable of complex communications by producing a very wide range of different signals for different purposes and for different audiences. These include adopting different postures and moving in different ways, but the most significant are their colour-changing abilities.

Cuttlefishes' camouflaging skills rival or even surpass that of chameleons. They have a number of different mechanisms for changing their appearance to match their surroundings. Their skin is covered in small papillary muscles that can contract to create rough lumps and bumps to mimic sharp, knobbly rocks almost instantly. They also can change colour extremely rapidly, and this is achieved by a number of different systems.

The skin's surface is densely covered in specialised structures called chromatophores. These are sacs containing grains of different coloured pigments – reds, yellows and browns – which are folded within muscles. When the muscles contract they open the chromatophore to reveal the pigments within. They have very localised control of these structures, creating detailed patterns by revealing different pigments in different parts of the body.

Below the chromatophores is a layer of cells that can disrupt the refraction and reflection of light, creating iridescent structural colours that appear metallic: blues and purples that look different from different angles. Below these is a layer of cells that, once uncovered, produce light colours, allowing camouflage against white backgrounds or producing white stripes contrasting against the dark pigments of the chromatophores.

Cuttlefish are masters of colour-change – both for camouflage and communication.

Aside from camouflage, these systems are put to heavy use in communication. They can produce a wide range of complex patterns instantly, which bring about different behavioural changes in the observer. Some of the patterns have been given names like 'flamboyant', 'weak zebra', 'intense zebra' and 'passing cloud'. This last one looks like blurry bands of different tone waving along their bodies. It is difficult to focus on, and is believed to be a tactic for distracting, confusing or even hypnotising prey, allowing the cuttlefish to strike.

These patterns are used to communicate aggression to rivals, readiness to mate to potential partners, or warning signals to possible predators – communicating with members of their own species as well as completely unrelated ones. Remarkably, they can have different patterns on their left and right sides; giving different signals to animals on either side of their bodies (for example a mate on one side, and a rival on the other).

Considering the complexity of their colour patterns, it is interesting that cuttlefishes are effectively colour blind. However they are able to discern different polarities of light, which is believed to allow them to detect the different tones in patterns as an alternative way to having colour vision. Because the structures that produce the iridescence and polarise light, cuttlefishes are believed to communicate with each other using patterns of polarised light that are invisible to many of their predators.

It's a shame that the process of preserving these amazingly coloured creatures strips all notion of pigment from them, so in museums they just appear matt white. This is one of the reasons that the Blaschkas made coloured models out of glass such as the next object (and these are fully explored in chapter 97).

58 Blue Sea Slug

BLASCHKA GLASS MODEL

Recycling Venom

I started this book by saying that evolution can only work with what it's got. Adaptations are usually honed by modifying a feature that the species already possesses. However, there is a way to cheat the system by acquiring the hard-earned (or more accurately, hard-evolved) characteristics of another species. In this way, an organism can benefit from a particular feature without actually developing it themselves. It is rather like those comic book characters that can absorb the powers of superheroes for themselves. This isn't to say that evolution isn't at work, as they still need to evolve the ability to take in and utilise the feature in question, and that is an adaptation in itself.

Nudibranchs are a group of about 2,300 shell-less carnivorous gastropod molluscs (the group which also contains slugs and snails) and are one of the groups that are often referred to as 'sea slugs'. They include an incredible array of brightly coloured animals from across world's seas in all habitats from the depths of the open ocean to rock pools. The vast majority live on the seabed, crawling around on their 'foot' like other gastropods, but the blue sea slug is an exception to this rule.

Blue sea slugs have a gas sac near their foot, which makes them float upside down at the water's surface. On this upward facing side (actually their anatomically 'bottom' side) they are a brilliant metallic blue, while their downward (anatomically 'top') side is pale. This is a classic example of countershading – a common phenomenon in marine species which makes them camouflaged from whatever angle they are observed in the three-dimensional space of the sea. Looking down from above, predatory birds will struggle to see them against the blue of the ocean, while looking up from below, fishes and other predators will find that these sea slugs' undersides are camouflaged against the brightness of the sky. Countershading can be seen in a huge range of swimming organisms, from penguins and cormorants to sharks and dolphins. It is a tactic used to camouflage both predators and prey.

Blue sea slugs are specialised to prey upon stinging cnidarians like by-the-wind sailors and Portuguese man o' war (chapter 68), which also have evolved gas sacs to make them float at the water's surface. These jellyfish-like animals use their stings to help them to immobilise prey, as well as a defence against potential predators. However, the sea slugs are not only immune to the stings of these creatures, but they have also evolved the ability to 'steal' the stings for themselves.

Cnidarian stinging cells – called cnidocytes – contain a structure called a nematocyst which holds the toxin as well as a trigger and a tiny hollow tube. When the cell is triggered by a combination of touch and the chemical detection of specific substances in the target animal's surface, the tube is fired at an incredible speed and force, injecting the venom into the victim's skin. It is a kind of mini explosion that means each cell can only be used once. As such, being venomous in this way comes at a significant energetic cost, and the blue sea slug's evolutionary knack of usurping the cnidarians' sting is a neat tactic.

The sea slugs eat the tentacles of the cnidarians, where the cnidocytes are concentrated, and somehow manage to avoid triggering the nematocysts as they make an incredible journey through their bodies. They segregate the stinging cells in their stomach, and divert them towards the finger-like frilly projections that stick out of their bodies. The nematocysts are then stored there in little sacs, waiting to be deployed against the sea slugs' own predators. Because they are effectively concentrating the stings of the already dangerous animals they eat, blue sea slugs can pack a sting that is very powerful.

This is not the only sea slug that acquires the adaptations of the species it eats. There are some species that sequester the algae that live within the corals that the sea slugs eat. The algae continue to photosynthesise inside their new 'home', and in so doing produce sugars that the sea slug benefits from. One species, the eastern emerald elysia, doesn't waste food by allowing the whole algal cell to live inside it, instead it just sucks out the chloroplasts (the green organelles where photosynthesis takes place in the algae) and stores them for its own gain.

The blue sea slug pictured opposite is made from glass, as historically soft-bodied creatures like sea slugs were very difficult to preserve. It is part of a collection of models made by the Blaschkas, who are now famous for intricately recreating anatomically perfect animals in glass (see chapter 97).

The eastern emerald elysia is a sea slug that removes photosynthetic chloroplasts from algal cells, benefiting from the sugar they produce.

59 Caterpillar With Parasitoid Fungus

DRIED SPECIMEN

Symbiosis: Zombie Death Grips And Parasite Manipulation

An animal's behaviour is controlled by a combination of nature and nurture: instinct and learning. And also zombie overlords.

There are some fungi that parasitize insects and modify their hosts' behaviour to meet their own needs, effectively turning them into zombies. This specimen is – at one end – a dead caterpillar (aptly a kind of ghost moth) and at the other a long rod of fungus belonging to the genus *Cordyceps*. This is the kind of object that museums struggle to catalogue properly – should it be entered on our systems as a fungus or a caterpillar?

Beginning by feeding on the internal organs of the caterpillar, the fungus slowly occupied its host's husk, until it was little more than a hard fungus sausage. It then grew the long stalk out into the air from which it would release its spores so that they might also land on a moth caterpillar, and start the whole gruesome cycle again.

The insect world is rife with parasitic and parasitoid fungi like this one (parasitoids are parasites that end up killing their hosts). Among them there are some incredible examples of the fungi altering the way that the

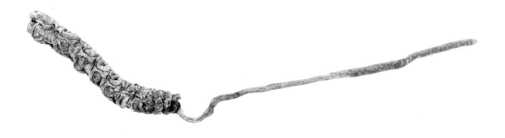

hosts behave, in order to maximise the chances of the fungus being able to grow its spores and infect other individuals with them. The process is called 'host manipulation'. It's an unpleasant way to be reminded of the fact that fungi are more closely related to animals than to plants.

A particularly striking example of host manipulation called 'summit disease' occurs across a wide range of insect groups including aphids, grasshoppers, butterflies and moths, beetles and ants. The fungus grows within the host's body, infecting various organs, but before it reaches a point when the animal can no longer move, somehow the fungus causes the host to climb to a high point like the top of a grass stem or tree and wait there to die.

This can be beneficial to the fungus as from this elevated position they can spread their spores far and wide on the breeze. Other versions of this kind of host manipulation cause the insects to go to a point where the temperature and humidity favour fungal growth, or where they are more likely to be encountered by other members of the hosts' species, increasing the chance of disease transmission.

The 'zombie control' can even specify the time of day that the hosts climb to their final resting places. In several species it has been shown that summit disease takes place most often in the late afternoon and evening. It is assumed that the humidity of the night is most advantageous to the fungus (rather than the potentially desiccating conditions of sunlight), as well as the likelihood of being transmitted to another victim.

Once in place at the tops of their twigs, grass blades or trees, it is in the fungi's interest to ensure that the dead or dying hosts do not fall off. The longer they stay in place, the longer they have to grow and spread their spores. This may also explain why the zombie insects are often found on the underside of leaves – as it decreases the chances of being spotted by predators or scavengers, which would not suit these fungi at all.

The fungi have evolved different ways of keeping their insect incubators in position. Most simply, the fungus can grow threads or roots to attach the insect to its substrate. However there are more shocking examples of host manipulation that bring about the same end. The 'death grip' is the most famous – infected grasshoppers, caddisflies and dung flies are all made to cling tightly with their legs locked around a holdfast. Other species are made to bite their substrate to stay in place. A number of species of ant, for example, are infected by *Ophiocordyceps* fungi, closely related to the one seen in this caterpillar specimen. This manipulates them to climb to the underside of a leaf and use their mandibles to clamp on irreversibly to a leaf vein, and wait there to die and sprout a sporing body. These ants were first found attached to leaves by Alfred Russel Wallace in Indonesia in 1859 (though there is no evidence that he attributed their behaviour to parasitic fungi). Incredibly, fossil leaves with the characteristic cut marks of zombie ants were found in 48-million-year-old sediments from Germany.

The most bizarre manipulative control to keep hosts in position has to be a species of grasshopper that when dying from a *Sorosporella* fungal infection is effectively glued to a leaf at both ends of its body with its own vomit and faeces, induced by the fungus.

Such parasitic manipulation is not restricted to fungi (it's even seen in humans infected with Toxoplasmosis). There is a species of nematode worm that parasitizes a species of Neotropical arboreal ant, *Cephalotes atratus*. An uninfected ant is all black, but infected ones grow a bright red abdomen so that it looks like the red berries of the trees the ant inhabits in an incredible example of parasite-induced mimicry. The ants' behaviour is also manipulated, to hold their red abdomens conspicuously in the air. This increases their chance of being spotted by berry-eating birds. Not only that, but the infected ants also show decreased anti-predator behaviour. The ants' abdomens are packed with nematode eggs, which once they are spread about in the dung of birds, are collected by the next set of victims, completing the cycle.

A 'zombie ant' that has been induced to climb on to the underside of a leaf by its *Ophiocordyceps* fungal parasite, and lock itself into place with a 'death grip'. The fungus has then sprouted a sporing body from the ant's head.

60 Bumblebee

Symbiosis: Mutualism – You Scratch My Back …

We are taught that evolution works by competition: 'survival of the fittest'. However, it isn't quite accurate. Species working together in harmony is one of the most powerful evolutionary processes in the history of animals.

Or in the history of life, in fact. At a cellular level, every plant, animal and fungus (among other things) is the result of cooperation between species. Mitochondria, the tiny organelles that power all of our cells; and chloroplasts, the green photosynthesising structures that produce sugar from sunlight for plants – and in so doing convert the world's carbon dioxide into breathable oxygen – are believed to be highly modified bacterial cells. Complex cells are thought to have originally derived by engulfing independent bacteria that were the ancestors of mitochondria and chloroplasts. Essentially this would mean that every animal and plant cell is a symbiotic collaboration between descendants of bacteria that live within it. There could hardly be two more important steps in the history of the world.

Symbiosis is the close biological interaction of two or more species. It comes in a number of forms – parasitism is an interaction where one species benefits and the other one is harmed. Mutualism is the cooperation of two species where both parties win. Commensalism (explored in the next chapter) is the form of symbiosis when one species benefits and the other is not significantly affected one way or the other.

Examples of mutualism are all around us. Lichens have always been held up as one of the most famous examples of mutualism – the happy symbiotic relationship of a fungus and an alga which has been written about for centuries. The fungi gain from the sugars the algae produce through photosynthesising, while the algae benefit from protection provided by the fungus. Amazingly, despite the fame of this interaction, it was only in 2016 that scientists discovered that there is actually a third species in the mix – a cyanobacterium. The three have evolved together to form a permanent relationship where none can survive without the other – this is the extreme end of the symbiotic scale.

Plants and pollinators like bumblebees have evolved hand in hand, each driving the diversity of the other.

Another famous example of mutualism is that of the clownfish and the anemone. Like jellyfish, anemones are members of the Cnidaria. They share with their free-swimming cousins the explosive cnidocyte stinging cells, both as a means of protection against predators and competitors, and to paralyse their prey. However, they are not completely immune to attack and a number of fishes and invertebrates are able to eat them given the chance. That's where the clownfish comes in; clownfish live among the tentacles and chase off the anemone's attackers. The anemone also benefits from the nutrients absorbed from clownfish poo, and leftover scraps from its meals.

In return, the clownfish gains protection against its own predators by hiding among the stinging tentacles. They have evolved a mucus coating over their bodies that stops them from being stung by the anemone, and from sticking to its predatory tentacles. The two have co-evolved in this relationship.

The fish and anemone story is charming enough, but it hasn't changed the world. Flowering plants, however, have had a hugely dramatic influence over life on earth, and to a great extent the history of their diversity is closely linked to a mutualistic relationship with pollinating insects.

A huge number of plant species rely on insects (and other pollinators, particularly bats and birds) for reproduction, and they have co-evolved with them over the earth's history. Many species passively spread their pollen on the wind from inconspicuous flowers, with their reproductive success dependent on some of it landing on the female part of a flower of the same species. However, plants that have more visible flowers rely on animals carrying their pollen from one flower to the next.

Species like this bumblebee have evolved hand-in-hand with flowering plants. The flowers have evolved an innumerable diversity of tactics to gain a pollinator's attention, and then to reward them for visiting. The shape, scent, colours and markings (including many that are invisible to human eyes) of flowers are for the pollinators' benefit. In turn the pollinators have co-evolved the necessary anatomy – including the shape of their mouthparts, legs and bodies – to reach the insides of the flower. Many plant species are dependent on a single, specifically adapted, species of pollinator.

The plant provides sugary nectar and rich pollen (more than the plant needs to reproduce) to feed its mutualistic partner. Bees have evolved little 'pockets' on their legs to fill with pollen to take back to feed the hive – these can often been seen as solid yellow balls on a working bee. As the bees forage on the flower, the fuzzy hairs on their thorax and abdomen become laden with a dusting of the sticky pollen. When they fly on to the next plant some of that pollen rubs off onto the stigma – the female part of the flower – allowing fertilisation to take place.

Bumblebee feet do a remarkable thing: they deposit scent as they feed on a flower. This lets other bees know that they have already drunk all of that flower's nectar. If you watch a bee forage, it doesn't visit every flower on a plant – this is because it is sensing the scent of other bees on the flowers that are 'empty' and doesn't waste time and energy on them. The scent decays over time, giving the flower a chance to replenish the nectar stocks before the next bee comes along.

In 2013, an even more surprising discovery was made – foraging bumblebees change the electric field of the flowers they visit. The rapid flapping of their wings in flight generates a positive electric charge (which helps pollen stick to it like hair to a statically charged balloon) and this is temporarily transferred to a flower when they feed. The study found that just as with scent-marking, bees avoided flowers with a positive charge as this suggested that another bee had been there recently.

The massive diversity of flowering plant species has evolved in large part as a result of mutualistic co-evolution with insects like this bumblebee. Likewise, the diversity of pollinating insects has arisen from their relationships with the different plant species. Neither could have happened without the other.

61 Sponge Crab

Symbiosis: Commensalism – Nature's One-Sided Relationships

As most humans know, not all relationships are evenly balanced. Sometimes one partner gives more and one gives less. Mutualism describes a relationship where both parties gain. Parasitism by contrast, is a pairing where there is a clear winner and a loser. Commensalism is a situation when one side benefits, while the other remains relatively passive – neither benefiting from the liaison nor suffering much inconvenience.

This specimen is made of two animals – a sponge and a sponge crab. It is a rather charming example of a commensal relationship. The story of this pair is quite similar to the more famous hermit crabs which use the empty shells of dead gastropod molluscs as a protective 'home' (this form of symbiosis, where one organism makes use of a suitable environment created by another who has died, is called metabiosis). Hermit crabs have undergone some fairly dramatic evolutionary changes, such as losing the hard protective exoskeleton from their abdomens, which has become slightly coiled, allowing them to squeeze into the spiral internal chamber of a shell. The advantage of this is that the shells are tougher than an average crab's carapace, and they can withdraw quite far inside them if attacked by a predator. They are also very good camouflage.

Camouflage is presumably the driving force behind the tactic used by the sponge crab, *Dromia vulgaris*. These crustaceans – and other members of their family – shape a piece of living sponge to fit around the curve of their carapace and live in it. Their anatomical modifications are slightly less dramatic than their hermit cousins – their back two pairs of legs are much shorter than the others and curve backwards over their bodies to hold the sponge in place.

Sponge and sponge crab then live together for the long term. Where the crab goes, the sponge goes. As the crab grows, the sponge grows. The crab benefits from the camouflage that the sponge provides (sponge crabs are unpopular aquarium species, as they can be very hard to spot). Sponges also carry unpleasant toxins, which presumably also benefit the sheltering crabs.

I feel like I should point out something of a museological nature on this specimen, although it may dent museums' reputations as scientific institutions. When preserved specimens are prepared for display in a glass jar, they are often fixed to a plate of glass to stop them from slumping to the bottom of the jar. Museum conservators (who prepare these specimens) often pride themselves on the subtlety of their skills by rendering their work invisible. They sew the specimen to the plate with ultra-fine thread and extremely discrete holes and knots. In this case, however, the conservator seems to be having a bit of fun (people who work in museums are human, after all): to complement the sponge 'hat', they have tied the crab into the jar with a dainty little beaded bow-tie. It's an important reminder that while natural history specimens are *natural*, they also display a bit of human interpretation and bias, from the hand of the person that prepared them

The definition of commensalism includes one partner – in this case the sponge – who does not significantly benefit *or* suffer from the symbiosis. In reality it's actually quite hard to determine whether the sponge is positively or negatively affected by its life of crab-harbouring. Sponges feed by filtering detritus from the water around them. In theory this means that they could be benefiting to some degree, as they can feed on the crabs' food scraps and faeces. Also the movement of the crab around the environment could facilitate water flow through the sponge, assisting its filter-feeding. As such, some people might classify this symbiosis as mutualistic. However, since the sponges can live perfectly well independent of the sponge crab (they are not called crab sponges, for instance), there can't be a significant advantage to being a crab's hat. In any case the benefit is very difficult to measure.

Other examples of commensalism carry similar potential problems: cost can be hard to calculate too. The oddly shaped remoras are fish with

incredible suction pads on their heads. They use them to temporarily attach to sharks' and whales' skin in order to hitch a free ride on their larger companions. This saves the remora from expending energy from swimming considerable distances as the bigger animals are doing all the work. This can be considered commensalism as the remoras are extremely streamlined and hundreds of times smaller than the animals they attach to, presumably meaning that the cost to the shark or whale is negligible. Again, however, it is very hard to measure whether the increased drag of having one – of sometimes many – remoras attached to their bodies could constitute parasitism.

There is a highly specific species of moth *Trisyntopa scatophaga* which is only found in a small area of northern Queensland, Australia. It lays its eggs in the entrances to the nest tunnels that golden-shouldered parrots build in termite mounds. The larvae are only known to feed on the dung of this single species of parrot. Not all nests have the moth, so while the cleaning up of dung could be helpful to the parrot, it can't be considered to be dependent on it. This means it's sensible to call it commensalism rather than mutualism, but again the line is blurry.

Golden-shouldered parrots are endangered. With any symbiosis in which one species is dependent on another, be it parasitic, mutualistic, or commensal, it is worth remembering that when that species dies out, so do those that were reliant on it. When campaigning to save charismatic species like the parrot, we shouldn't be forgetting the moth.

The suction pads on the heads of remoras allow them to hitch a ride on larger animals like sharks and whales.

62 Common Cuckoo

Brood Parasites: An Evolutionary Arms Race

Evolutionary biologists talk about competition, but can organisms ever *win*? If the success of one individual comes at a cost to another, there will be evolutionary pressure for the 'loser' to fight back. Take a theoretical example: animals evolve the ability to digest plants, so the plants suffer. This puts pressure on the plants to evolve toxins to deter the herbivores; but they then evolve the mechanism to break down the poisons. The plants evolve stronger toxins, and the herbivores stronger tolerances, and so on. Nature is full of evolutionary arms races, where there is one side developing adaptations to exploit the other, which in turn is driven to evolve counter-adaptations in an ever escalating cycle.

Everywhere we look these arms races are apparent – weapons of attack drive the evolution of weapons of defence, and *vice versa*. Visual acuity drives the evolution of camouflage, and *vice versa*. Metabolising toxins

drives the evolution of stronger toxins, and *vice versa*. It is the *vice versa* that matters – the one-upmanship is constant.

When we think of parasitism it is often tiny worms, mites and insects living on much larger hosts that come to mind, but sometimes the parasites are bigger than the species they are parasitizing. Brood parasites are species that avoid the costs of raising their young by 'dumping' them on somebody else.

Most famously, cuckoos are brood parasites. When you break it down, what cuckoos do to their hosts is both ingenious and – to our human sensibilities – ghastly. They wait hidden near a potential host's nest for them to leave briefly to forage, and swoop in and lay one of their own eggs – to avoid detection cuckoos can do this in under ten seconds. Birds can count, so before doing so they steal one of the host's own eggs, which they later eat. The intention is that the host bird will then incubate the cuckoo's egg as if it were one of her own.

Cuckoos develop their eggs in their own bodies for a day longer than other similarly sized birds, which means that they often take less time to hatch than their hosts' eggs. This gives the chicks a window of opportunity to kill all of the hosts' young before they hatch. Blind and featherless, a newly hatched cuckoo scoops each of the host's eggs between its shoulders and backs them out of the nest. If it happens to hatch after the host's chicks, it doesn't matter – it throws them out of the nest too.

Having removed all of the competition for their foster parents' attention, they then beg their hosts to feed them constantly until they are ready to fledge. This can be quite an absurd sight, as the species that cuckoos parasitize are significantly smaller than them. Reed warblers, for example are about 13cm long, while cuckoos are over 30cm, and much stouter.

The cost to the host's reproductive success for a parasitized clutch is absolute: they do not pass on any genes because the cuckoo has killed all their young. With such a high price for 'losing', we would expect that host species would evolve measures to detect cuckoos in the nest. And indeed they do.

Most host species have evolved the ability to discern a cuckoo's egg from their own, and reject it. Common cuckoos parasitize a range of different species, and amazingly have evolved the ability to mimic a range of very different eggs. Cuckoos that lay in pipit nests lay brown spotted eggs, like a pipit; those that parasitize reed warblers lay green spotted eggs, like a reed warbler; and so on. It is an evolutionary arms race to lay an egg that best mimics the hosts' eggs, and for the host to be able to tell the difference.

Because hosts are constantly evolving improved abilities to spot a cuckoo's egg, the cuckoos have to get better at mimicking them. What's remarkable is that all the different egg types are laid by the same species of cuckoo, but each female is a specialist to one kind of host. It is passed

A tiny adult redstart feeding a large cuckoo chick.

down in the maternal genes, so that they parasitize the same species of host as they were raised by. They don't ever split into new species as the males don't discern between them, so they keep mixing the genes up.

Interestingly, in experiments using model cuckoo eggs, some host species which are only rarely parasitized reject over 80 per cent of mimicking cuckoo eggs from their nests. This suggests that they may have 'won' the arms race – perhaps the reason that they are now only rarely parasitized is because cuckoos have been forced to move on to less discerning species. By contrast dunnocks never reject cuckoo eggs, even though they do not mimic their own: perhaps the dunnock is a 'new' host, and has not had the chance to evolve the ability to detect a parasite yet. The host species that are the most discerning are those that the cuckoo eggs mimic the most closely.

It would seem that host species learn what their own eggs look like when they breed for the first time. This is one reason why discernment can never be perfect, as they may be learning from a clutch that has a cuckoo egg in it. It's also the likely reason that hosts haven't evolved the ability to recognise a giant cuckoo chick in their nest. If they have a cuckoo in their first batch, and were to 'learn' that this is what their own chicks look like, they would end up rejecting all their own young from subsequent clutches, as they didn't look like a cuckoo.

63 Horseshoe Crab

PRESERVED SPECIMEN

'Living Fossils'

There are no such things as living fossils, despite what you may have been told. It is a term that is widely used for species that either have a long fossil record with apparently little change, or those that have been discovered alive after having only been known from the fossil record (they had previously been assumed to be long extinct). These are unscientific

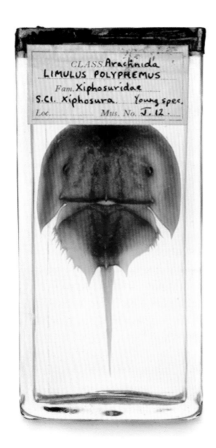

ways to describe organisms, and they do not stand up to scrutiny.

Coelacanths (chapter 24) are arguably the most famous 'living fossils'. These fish (a kind of lobe-finned fish, related to the fishes that evolved the ability to live on land and become our ancestors) have a long and diverse fossil record. Based on fossil evidence, palaeontologists assumed that they became extinct 66 million years ago after a prosperous career spanning over 300 million years. They had been studied as fossils for around a century before a live specimen was caught in 1938 in the Indian Ocean. This is how they earned the moniker 'living fossil'.

What's problematic in this instance is that the pseudoscientific term 'living fossil' is dependent on the historical order of events that scientists happened to make discoveries. If they had known about the living species before the fossils were described, they would not be called living fossils. It is an accident of human history completely independent of coelacanth biology.

Horseshoe crabs are also regularly referred to as 'living fossils'. The argument goes that fossil horseshoe crabs have been around for hundreds of millions of years, and over that time they don't seem to have done much evolving: fossils from 445 million years ago appear roughly similar to the four species alive today. This reasoning doesn't hold up to examination either.

To start with, it's impossible. There is no way that a lineage would have gone for hundreds of millions of years without evolving. Sure – the superficial similarity of the carapace of a horseshoe crab from the Ordovician may resemble a modern one, but that doesn't mean they are the same. These assumptions are often based on observations of the external features, and tell us nothing of their internal anatomy (which doesn't fossilise). It is of course impossible to test whether modern and fossil horseshoe crabs could have interbred, but the process of genetic drift (the non-directional evolutionary changes) means that their genes would be very different.

Perhaps more important than demonstrating that it *couldn't* be true, however, is to show that it simply *isn't* true: fossil horseshoe crabs are morphologically different to modern ones. This is how palaeontologists have managed to describe at least ten different genera containing many more species. In fact, modern species of horseshoe crabs don't appear in the fossil record at all. They have evolved.

Some of the differences between the fossil species are slight – as they are between most members of the same family. Horseshoe crabs have four pairs of walking legs with little crab-like pincers on the end, and one pair with little scoops on to help them burrow. This was thought to be the standard horseshoe crab way of doing things today and through the fossil record.

However in 2012 a 425-million-year-old specimen from Herefordshire in England was described with five pairs of legs which each had two branches (appearing like ten pairs). It was called *Dibasterium durgae*, and durgae refers to the many-armed Hindu goddess Durga. It would seem that this is the form that the ancestors of modern arthropods took, as there are fossils with similarly branched legs from the early stages of insect evolution too. At some point in the early stages of horseshoe crab evolution they lost one half of each pair, arriving at the version we see today.

One of the first facts often provided about horseshoe crabs is that they are not actually crabs. Instead, they are more closely related to the spiders, scorpions, mites and sea scorpions, making a group called the chelicerates. This taxonomic discovery was made in the Grant Museum of Zoology by our third curator E. Ray Lankester (in post from 1875 to 1891), possibly using the specimen shown here. One of his lines of evidence was that all of these groups share structures called chelicerae, used for guiding

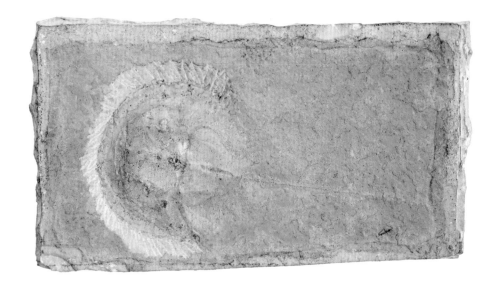

The 155-million-year-old Jurassic horseshoe crab *Mesolimulus walchi* from Solnhofen in Germany.

food into their mouths. They look a bit like mini legs: they are jointed and in horseshoe crabs have small pincers at the end.

It has since been demonstrated that chelicerae in chelicerates correspond evolutionarily to antennae in other arthropods like crustaceans and insects. Interestingly the strange leggy fossil horseshoe crab *Dibasterium* had elongated and apparently flexible chelicerae – more like antennae.

It is clear that there have been significant changes in the fossil history of horseshoe crabs, but since *Dibasterium* they haven't changed a huge amount. This doesn't justify the name 'living fossils', however. It's not well understood why they have been relatively conservative in their anatomy, but it doesn't make sense to *expect* large-scale changes over time; even over hundreds of millions of years. The rule 'if it ain't broke, don't fix it' would seem to apply. The habitats that horseshoe crabs live in haven't changed. Once evolution produced something that was well adapted to that habitat early in the group's history, what pressures would there be to change it into something else?

More than anything, however, maybe the term 'living fossil' persists so vehemently because some animals just look a bit prehistoric, which isn't a particularly scientific way of naming animals.

64 New Zealand Native Frog

PRESERVED SPECIMEN

Evolutionary Distinctiveness

There are only four species of frog native to New Zealand and they are unfortunately also tarred with the 'living fossil' brush because of their limited geographical range, poor jumping ability, and seemingly unspecialised skeletons that resemble those of fossil frogs from the Jurassic.

As I have mentioned, no *species* alive today should be considered primitive as they are just as 'evolved' as any other organism that has got this far. However, these four rare New Zealand native frogs do have a number of seemingly primitive *features*. Frogs' closest relatives are the salamanders, and some of the New Zealand frogs' features indicate their relationship to them. One of the defining characteristics of all frogs is that they do not have tails – indeed the name for the frog order is Anura, which means 'without tail'. Therefore by definition New Zealand native frogs do not have tails as they belong to this order, but they do still have tail *muscles*: the caudalipuboischiotibialis muscles. New Zealand native frogs have the means to wag their tails, but there is no tail to wag.

Other frogs have reduced the number of vertebrae as an adaptation to efficient jumping (fewer joints means less loss of elastic energy through the body) – they have eight or fewer. By contrast New Zealand frogs have nine. This feature is only shared with their closest relatives – the tailed frogs of North America (named for the shape of the male reproductive organs). Together these two groups make up the most ancient family of living frogs, the Leiopelmatidae.

These frogs show other unique features – they have round pupils and rods of cartilage in their abdomen (which has given the New Zealand frogs the occasional name of 'ribbed frogs' – other frogs do not have any ribs). They don't have eardrums – a significant feature of most frogs' heads. This is probably related to the fact that they can't croak as they haven't developed vocal sacs: they lack the froggy ability to inflate their throats. It is believed that they communicate more like salamanders, using chemical signals to recognize other individuals.

New Zealand native frogs spend most of their life on land under stones and logs in moist environments. Other frogs evolved advanced swimming techniques after this group split from the rest of the frog family tree. When they do swim, ribbed frogs do it differently to their cousins – they use alternate leg kicks rather than the symmetrical 'breast stroke' that's typical of frogs. This is less energy efficient. They don't have webbed feet, and their hop is rather graceless – they belly flop at each landing, having to right themselves before jumping again.

Interestingly they don't really have a tadpole stage – they emerge as tailed froglets directly from the egg (which in three of the four species isn't laid in water), and then they climb on their father's back where they stay for several weeks until they have completely metamorphosed. New Zealand frogs don't flick their tongues when feeding – they just bite their invertebrate prey as they forage in the leaf litter and rocks, but they don't seem to venture far. Research on one species – the Maud Island frog – found that they tended to stay within a 5m radius for years at a time. They are also very long lived, with individuals recaptured over 30 years.

The four species of New Zealand native frog (Archer's frog, Hochstetter's frog, Hamilton's frog and the Maud Island frog), along with their North American relatives, represent the most ancient frogs in the world. Their ancestors started on a separate evolutionary journey from all other frog and toad lineages over 200 million years ago. Once they had split away on their own branch of the evolutionary tree, their story became distinct.

Evolutionary distinctiveness is a measure of how unique a species is and it's a really interesting way of thinking about animals. The plight of tigers, for example, gets a huge amount of media coverage. I'm lucky enough to have seen tigers many times in the wild, and each time it is genuinely an emotional experience. It makes me stop breathing. Such a reaction is not common when people encounter a frog.

If tigers were to die out it would be genuinely devastating for a number of reasons (not least because of the role they play in their ecosystems), but we wouldn't have really lost much from an evolutionary point of view. There are species that are quite a lot like tigers elsewhere in the world, so we wouldn't have lost a significant evolutionary unit. Nearly all of the genes that make up tigers would survive in other members of the cat family. Tigers are not very evolutionarily distinct.

Leiopelmatid frogs, on the other hand, are extraordinarily distinct. Because of the many millions of years since they diverged from other frogs, they hold a relatively high proportion of genetic uniqueness. The Zoological Society of London have developed a means of prioritising which species should receive funding for conservation (the sad truth is that there is absolutely not enough money to save them all). It is called the EDGE programme, standing for Evolutionarily Distinct and Globally Endangered. It ranks species by combining a measure of their rarity with the amount of close relatives it has. Among the amphibians, Archer's frog scores in a Number One. It is the species that should be prioritised higher than any other. Their population currently number in the hundreds, and they have so few living relatives.

Evolutionary distinctiveness as a measure of inherent conservation value makes sense if you think of a hypothetical evolutionary tree. If you lose a twig (like a tiger), the tree doesn't suffer much. But if you lose a whole branch (like a ribbed frog), it is much more significant. The branch that has New Zealand native frogs at its tip stems from deep on the frog-tree's trunk. It has been growing for 200 million years, but over the last few decades it has been withering.

65 Kangaroo

PRESERVED JOEY ON A TEAT

Conveyor Belt Reproduction

A newborn kangaroo is an incredible thing. Although kangaroos grow to 85kg in weight and over 2m long, they are born the size and shape of a jelly bean, weighing 0.8g. This jelly bean, however, despite limited internal development, has fully functioning forelimbs and lips, which it uses to make the long climb largely unaided from the birth canal through the mother's fur uphill to the pouch, whereupon it attaches to a teat. At the other end of the marsupial scale, the honey possum produces the smallest baby of any mammal: newborns are an incomprehensible *5mg* – about the quarter the mass of a grain of rice.

As such, marsupial reproduction is characterised by a very brief period of foetal development in the womb, followed by a long period of development suckling milk on the teat. This is in stark contrast to the more popular method (numerically speaking), which is practised by us placental mammals. Placentals have a long period of development in the womb, which is finished by a short period of development suckling milk on the teat. In the case of eastern grey kangaroos, the young spends just thirty-six days in the womb, then spends eight months without leaving the pouch at all. For the next three months it will venture out to explore the world outside, but regularly climbs back in. It permanently moves out at

11 months but continues to suckle (by sticking its head in from outside) until it's around 18 months old.

One of the earliest written accounts of the kangaroo was by Lieutenant Watkin Tench who travelled with the First Fleet, which established the British Australian colony at Sydney Cove in 1788. He said that kangaroo reproduction was 'contrary to the general laws of nature',* and in so doing cast aspersions on the marsupial way of doing things. He was amazed that such large kangaroos grew from the tiny joeys he found in their pouches (and to be fair, that *is* amazing). Tench wrote that the joey 'descends from the belly into the pouch by one of the teats' – he thought that they emerged directly from the nipple rather than the vagina.

This misconception started with the very first description of an Australian mammal by a European. Contrary to popular belief, Captain Cook did not 'discover' Australia in 1770. For one thing, such ideas outrageously dismiss the fact that Aboriginal Australians colonised the continent at least 50,000 years ago. But Cook wasn't even the first *European* to reach Australia – Dutch, Portuguese and Spanish traders – among others – had been navigating the west coast for at least 200 years before Cook. Dutchman Francisco Pelsaert encountered tammar wallabies on the Houtman Abrolhos islands, 80km west of modern day Geraldton, Western Australia in 1629. Despite having survived a horrific shipwreck and mutinous bloodbath among the crew of the *Batavia*, he still took the time to describe the wildlife:

> there are a large number of Cats [wallabies] ... Their generation or procreation is Very Miraculous, Yea, worthy to note: under the belly the females have a pouch into which one can put a hand, and in that she has her nipples, where we have discovered that in there their Young Grow with the nipple in mouth, and we have found lying in it some Which were only as large as a bean, but found the limbs of the small beast to be entirely in proportion, so that it is certain that they grow there out of the nipple.**

It was not until 1830 that the marsupial birth and crawl to the pouch was correctly described, by Alexander Collie, setting the record straight.

---

* Tench, W., *A Complete Account of the Settlement at Port Jackson in New South Wales. Including an Accurate Description of the Situation of the Colony; of the Natives; and of its Natural Productions* (G. Nicol and J. Sewell: London, 1793)

** Drake-Brockman, H., with all translations from the Dutch and Old Dutch by Drok. E.D., *Voyage to Disaster: The life of Francisco Pelsaert, covering his Indian report to the Dutch East India company and the wreck of the ship Batavia in 1629 off the coast of Western Australia together with the full text of his journals concerning the rescue voyages, the mutiny on the Abrolhos Islands and the subsequent trials of the mutineers* (Angus and Robertson: Sydney, 1963)

The denigration of marsupials as something inferior, 'contrary to the laws of nature' was perpetuated by Tench at the foundation of the Australian colony, but it has been pervasive (see the striped possum, chapter 40). It was a very long held view that marsupial reproduction was primitive, and altogether not as 'good' as the 'normal' way of doing things among us placental mammals. Indeed the taxonomic term for placental mammals, Eutheria translates as 'true beasts', while the marsupial term Metatheria means 'half beasts', underscoring traditional views towards them.

In fact pouch development is not evolutionarily primitive. It is an extremely sensible system in an unstable environment, such as Australia. If drought strikes and a mother can no longer feed both herself and her young, abandoning the baby could save her own life. This is much easier to do by reaching into a pouch and pulling the baby out than resorbing a large foetus in the womb. Marsupials give themselves the chance of surviving if things turn south, without having invested much in a young developing joey. As joeys get older, the energetic investment is greater and the likelihood of abandonment decreases.

Despite Pelsaert's mistake, kangaroo reproduction *is* very miraculous (or at least an amazing result of millions of years of evolution). Once a young kangaroo has ventured out of the pouch to begin exploring the world (referred to as a 'young-at-foot'), they still stick their heads in to suckle. Nevertheless this means there is room in the pouch for another occupant. Kangaroos can have a fertilised egg waiting in a state of arrested development for the right time to give birth (a system called embryonic diapause). Once the big joey frees up space in pouch, a new jellybean is born and can climb in. It will attach to a different nipple to the one the young-at-foot uses, and that nipple will produce different milk. Kangaroos can provide differently constituted milks to suit differently aged joeys at the same time.

Indeed they can have a fertilised egg in stasis in the womb, a tiny young permanently attached to the nipple in the pouch, and a larger young-at-foot still suckling all at the same time. If conditions are good they can run a constant conveyor belt of reproduction in this way. There's nothing inferior about that.

66 Koala

Evolutionary Constraints

Every zoologist has their own favourite group of animals, and mine is marsupials. As I've mentioned, this group sometimes suffers criticism from the more numerous type of zoologist who studies placental mammals. They say marsupials are boring, stupid, primitive, too few in number to care about and are altogether inferior to 'normal' mammals. They are wrong.

Whenever I go to Australia to undertake ecological fieldwork I am struck by the diversity of the mammals there. You can travel 200km and find, for example, a different species of small carnivorous marsupial doing a similar thing to the one you saw the day before, only in a slightly different environment. Go another 200km and you could find a third.

However, the three species would look pretty similar. One of the major downsides of marsupials, from a biodiversity point of view, is that they haven't evolved the range of forms that placental mammals have. While there is one semi-aquatic species of marsupial – the yapok – it could hardly be compared with a whale or a seal; there are gliding marsupials, but they can't do what

Marsupial hands do not show the anatomical diversity seen in other mammal groups, as they are needed to climb into the pouch.

bats can do. Marsupials and placentals have both been evolving for the same length of time – probably 160 million years; why did flying, swimming or even galloping never arise in marsupials?

It would appear it is down to their method of reproduction.

Mammals produce babies in one of three ways. The five species of monotreme (platypuses and echidnas) lay eggs. The 335 species of marsupial (e.g. kangaroos, quolls, bandicoots, opossums and wombats) give birth to highly under-developed (altricial) young that do most of their growing by suckling milk on a teat. The 5,100 species of placental (e.g. dogs, primates, rodents and nearly everything else) give birth to well-developed (precocial) young who finish off development with a short period of suckling. What is it about their strategy that has limited marsupial diversity so much?

When marsupials are born, which is after just a few weeks in the womb, they look like a bean with massive lips and strong arms. That makes sense if you have to climb up to a pouch and attach onto your mother's teat and suckle milk for months. Using some cutting-edge technology – laser scanners, micro-CT scanners, 3D digitisations and 3D microscopes, as well as good old-fashioned microscopy – biologists including Professor Anjali Goswami at University College London and her colleagues have studied the morphology of developing young. Their work shows that newborn marsupials have well developed facial and forelimb skeletons and a very basic nervous system – the bare minimum for that first expedition from the vagina to the pouch. In contrast placentals develop a complex nervous system earlier, and limbs come later as they aren't needed early on.

The need to physically climb through the mother's fur, arm over arm, and then attach to a teat by the lips has placed a significant constraint on the ability of marsupials to evolve the range of diverse body shapes we see in placental mammals. Given that climbing arms are so important in their first moments, they can't be changed later in life to a wing like a bat, a fin like a whale or a hoof like a horse.

Actually, hoof-like structures did evolve in one marsupial lineage – the bandicoots. Bandicoots are pointy-nosed omnivores shaped a little bit like a rabbit. They do not climb into the pouch: the mother positions herself to allow the newborn 'jelly beans' to slide down a moistened path into a rear-facing pouch. This means they have no need to crawl, which freed up the evolutionary constraint on their arms. Low and behold, the pig-footed bandicoot evolved, with hoof-like feet. Tragically, however, the activities of Europeans drove it to extinction in the 1950s, so we will never know where this unique marsupial evolutionary trajectory may have led. In placentals, of course, hooves are so successful an evolutionary development that hoofed mammals have developed into a massive diversity of species

A young opossum. Marsupials are born at an early stage of development, but have well-formed arms and lips for climbing and suckling in the pouch.

from hippos and camels to mouse deer and giraffes, not to mention that they are the ancestors of whales and dolphins.

Goswami's studies also found that the diversity of forms seen in the skeleton of the marsupial mouth is similarly restricted when compared to placentals, because of the importance of suckling for such a long phase in their development. Placental mammals have evolved a higher diversity of facial shapes, while marsupials have been relatively conservative in what they have evolved.

It can be tempting to think of evolution as a process that is constantly promoting change, but most of the time it is a stabilising force, stopping changes that don't 'work' from becoming established. Likewise, there are many features of organisms that are limited in how much they can change, because of the crucial role they play in development. These constrain the paths that evolution can take a lineage down. In marsupials, the need to be able to climb to the pouch and suckle is so strong that it limits the scope for evolution to change things.

67 Cane Toad

PRESERVED SPECIMEN

Biological Control

As a scientist, with Vulcan-like levelheadedness, my outlook on the natural world is totally free of emotion. My interactions with it are purely perfunctory, in order to amass and analyse cold data, motivated solely by the advancement of scientific understanding of solid facts. The natural world is only there to be databased. It is irrelevant whether facts are 'interesting' or not, all that matters is if they are useful for detecting some larger pattern. Anyone who says otherwise is a panda-hugging sentimental fluff-monger ...

Wouldn't it be weird if ecologists thought like that? On the one hand science is supposed to be independent of emotion, but on the other most of us are only in it because of our emotional attachment to the subject matter (animals, plants and ecosystems).

Throughout this book I have taken the chance to rave about the animals that amaze and excite me. Here I'm going to highlight one that I utterly despise. This is a strong word but I stand by it. You would be right to say it's not the animal's fault. I know this and hold people entirely to blame for the damage this

Toxic and predatory, cane toads have been agents of destruction since their deliberate and misguided introduction to Australia.

species does. That doesn't stop me getting irked when I see them in a habitat that they are destroying.

Sometimes species are introduced from one region to another, and in their new homes they find that there are no predators or parasites that can impact on them, so the population explodes. An option to control them may be to look to the pests' native range for a predator or parasite that it had evolved alongside and introduce that to the new habitat too, in the hope that it dampens their numbers.

This is called biological control, and historically there are many examples of it going terribly badly. The second species, introduced to control the first, can go on to cause terrible ecological damage as well. It's the environmental equivalent of the old woman who swallowed a fly. The only cases when biological control has worked have been extremely well researched and piloted, to determine that the controlling species doesn't represent a threat in itself. This is not what happened with cane toads in Australia.

Cane toads are a large species of amphibian, native to Central and South America from Texas south to the Amazon and Peru. What they are famous for, however, is being one of the world's 'worst' invasive species. They are major pests in Australia, the Philippines, the Caribbean and Hawaii.

A total of 101 cane toads were deliberately introduced into Australia (from a feral population in Hawaii) in 1935 by The Australian Bureau of Sugar Experimental Stations. It was a disastrous ecological mistake. The intended reason for the introduction was to control 'cane grubs' – beetle larvae which were damaging Queensland's sugar cane crops. This is an odd example of biological control where the control species wasn't actually a known predator of the pest species. It was a catastrophic stab in the dark.

This error is unbelievable as the grubs live on the root stock (underground) and the adult beetles live in the cane plants (up off the ground). Whilst the toads are voracious predators, they hunt *on the ground* – not under it or above it, so there was no chance this experiment was going to work. Indeed it didn't – cane toads had no effect on controlling cane grubs. They really should have a different name.

What they did succeed in doing is eating vast numbers of animals that are smaller than them – invertebrates, native frogs, lizards, small mammals, birds ... but not cane beetles. Cane toads can reach 30cm in length, so most things in Australia are much smaller than them. Big toads are 1kg in weight, and if that weren't awful enough I've heard reports of toads getting up to 3.4kg. Human babies start smaller than that.

As well as decimating the small animals, they also kill many of the big ones. The giant lumps behind the toads' eyes are called parotid glands, and they are full of poison. When distressed (for example, by being eaten), the glands ooze a sticky milky bufo-toxin which is more than capable of killing animals up to the size of a dog. That includes nearly all of Australia's predators – dingoes, quolls and other marsupial carnivores, crocodiles, native rodents, freshwater turtles, big fishes, raptors, owls, monitor lizards and snakes. Not having evolved alongside them, these animals have no natural immunity to the toads' toxins.

One has to admire cane toads' adaptability. Since their arrival in Gordonvale near Cairns in far north-eastern Australia, toads have rapidly spread south and west, and are now well into Western Australia and New South Wales. Recent research has found that toads on 'the invasion front' have evolved stronger and longer limbs, allowing them to disperse further and quicker. I've done fieldwork in the Top End of the Northern Territory where nearly all of the mammals had disappeared, but we were seeing *hundreds* of toads each night. I've also done a lot of fieldwork in the Kimberley in Australia's far northwest prior to the toads' arrival. By the time this book is published they will have arrived at these study sites and that is truly terrifying as it is currently extremely rich in small mammals, frogs, snakes, crocodiles and lizards.

Over recent decades ecologists have sought a 'cure' for the scourge of toads in Australia, but as they march ever westward and southward in a wave of destruction success stories are extremely difficult to find. Current research is investigating the deployment of young toads ahead of the invasion front, which only carry enough toxin to make a predator sick rather than kill it, to 'train' monitor lizards not to eat larger toads. Similarly, sausages are being made from toad meat and either small amounts of their skin (to give a low dose of poison) or a nausea-inducing chemical. By scattering these around habitats the hope is that quolls will eat them and learn to associate the smell and taste of toad with becoming ill, and avoid them. Both projects are showing signs of success for these species, but they will not save the habitats from the march of the toad.

A northern quoll – one of the species of marsupial carnivores threatened by invasive cane toads.

68 Portuguese Man O' War
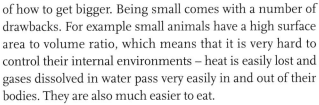
PRESERVED SPECIMEN

Colonial Animals

The earliest animals in the fossil record were tiny – mere millimetres across. One of the greatest steps in the evolution of life was to solve the question of how to get bigger. Being small comes with a number of drawbacks. For example small animals have a high surface area to volume ratio, which means that it is very hard to control their internal environments – heat is easily lost and gases dissolved in water pass very easily in and out of their bodies. They are also much easier to eat.

However, growing bigger comes with the problem of structural support – large objects need some form of stabilising skeleton to maintain their shape. In the vast majority of non-microscopic (macroscopic) animals this is done in one of three ways: a hard internal skeleton, a hard external skeleton, or a fluid skeleton that uses the hydro-static pressure of contained liquids to hold a shape.

There is a way to get round the constraint of needing such a skeleton, and that's to become colonial. Colonial animals are formed of a number of small genetically identical individuals fusing together to act like a single large animal. Instead of growing by increasing in size they grow by increasing in number. Effectively they are modular – the whole large animal is formed by replicating lots of small individual modules. The juvenile single animal grows only a little before reproducing asexually by budding off an identical copy of itself, but instead of splitting to form a new animal, the two individuals remain attached. They both then bud so there are four of them, then eight, and it keeps going until very many individuals are stuck together in a huge colony of hundreds or thousands.

Colonial lifestyles have evolved a number of times in invertebrates, including in the sponges and bryozoans, but it is the cnidarians who are the masters of it. This diverse phylum includes jellyfish, anemones and corals. Within this one group living colonially has evolved independently among the corals, sea pens and hydrozoans, but none have done it as dramatically as siphonophores like the Portuguese man o' war.

This species looks very much like a jellyfish, but it is not. It has a mass of stinging tentacles hanging from a central 'bell', but actually it is formed of thousands of tiny individual animals called zooids working together in a single colony to behave like one large animal. It raises the question of what it means to be an individual. Many colonial cnidarians that live fixed to one spot, such as corals, are formed of morphologically identical zooids, but siphonophores are an example of a cnidarian that has formed colonies of individual zooids with *different* specialised morphologies and functions. There are man o' war zooids that are responsible for hunting, digestion and reproduction, for example. If separated from the others they cannot survive on their own: they are each dependent on each other.

Their tentacles can reach 20–30m long, so if two different tentacles spread horizontally in different directions from the 'body', they could be considered the longest animals on earth.

Cnidarian bodies typically come in two forms – free-floating medusae (singular medusa, typified by adult jellyfish) and sedentary polyps (typified by adult anemones). Most species include both forms to varying degrees at some point during their life cycle. Man o' wars are unusual in that they include medusae and polyps at the same time.

All the zooids in a man o' war are descended from a single fertilized egg and so they are all genetically identical. The egg develops a polyp that buds to give rise to all the other zooids in the colony. The bit of the polyp that in other species holds it to the sea floor expands and folds to form a gas-filled float called a pneumatophore. Some zooids that bud off develop to collect prey using tentacles with stinging cnidocytes, others do the digesting and reproducing.

The long tentacles fire their venom-filled nematocysts when they come into contact with small fishes, squid and other invertebrates, which either paralyse or kill them. Tentacle cells then contract to drag the prey into range of the digestive polyps, which surround the food and release enzymes to break it down into nutrients that are then absorbed by all the zooids.

In this way the individuals work together to feed the whole colony. They behave in a way that allows these tiny animals to beat the constraints of their smallness, and become significant players in their ecosystems.

69 Mantis Shrimp

PRESERVED SPECIMEN

Predation

One of the features that all animals share is that they derive their energy from eating other organisms. Some have evolved to digest plants, some consume rotting material, some filter particles of organic matter from the environment, and some eat other animals. One order of crustaceans arguably has the most spectacular predatory adaptations of all: the mantis shrimps.

Mantis shrimps are most famed for the way that they disable their prey using the incredible feats of their second pair of legs. The 450 or so species in the group can roughly be categorised into two groups – spearers and smashers – depending on the way these appendages work.

The specimen overleaf is a spearer: using the elongated, barbed spears on their second 'thoracopod' (calling it a leg isn't quite right, as they aren't used for walking), species like this one spear their prey in lightning quick attacks. They are ambush predators that sit in wait, hiding in sand burrows or rock crevices for an appropriately sized animal to pass by, then they pop out and rapidly stab it on the end of their mantis-like limbs.

Smashers, on the other hand, have modified their thoracopods into extremely well strengthened hammers by thickening their exoskeleton on what roughly equates to their elbows. Rather than being ambush predators, they hunt more actively than spearers for more sedentary prey such as shelled molluscs, crabs or other crustaceans. Using their 'hammers' they can punch so hard that they smash through the solid casings of clam shells and crab carapaces to access the meat inside. As a result, larger species have become notoriously challenging to keep in aquaria as they are capable of punching through the glass. They are also known as 'thumb-busters' for the damage they can inflict on those who handle them.

The hammers of smasher mantis shrimps can reach incredible speeds of 23m per second, which is equivalent to 80km/h – it is the fastest punch on earth. This is even more impressive considering it is achieved in the viscous environment of sea water. The acceleration is faster than a .22 calibre bullet.

These speeds and forces are far greater than can be achieved by muscle contraction alone, and mantis shrimps have employed a number of physical power amplification devices to achieve them. They are they only animals known to employ saddle-shaped springs – a common form in architecture used to evenly absorb high pressures. In addition to these springs there are ratchets, latches and elastic structures that are combined to allow their limbs to quickly store up energy before being released in their high-speed blows.

Super high-speed cameras have shown that mantis shrimp punch with such force that the water trapped between their limbs and their prey

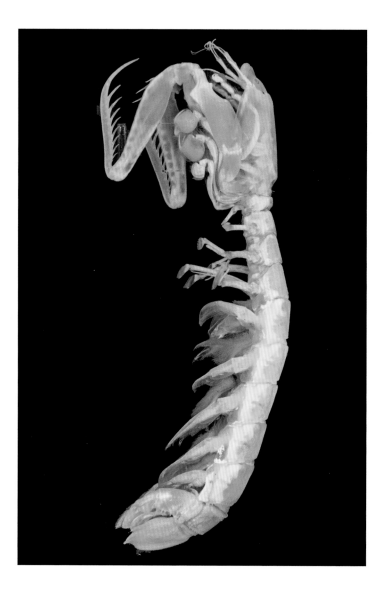

suddenly vaporises. This forms what are called cavitation bubbles, which themselves are highly destructive. They instantly collapse, causing a mini-explosion, the force of which deals a second blow to the victim. Cavitation bubbles that form when solid metal boat propellers cut through the water at high speed can blow holes in the propeller. When mantis shrimp perform a one-two punch on their prey, really they are getting hit with a quadruple stroke from the collapse of the two sets of cavitation bubbles.

Mantis shrimp eyes are equally intriguing. They have the largest known number of photoreceptor types in any animal. These are the cells that respond to different wavelengths of light. We have three – very roughly corresponding to red, green and blue light. By comparing the level of absorption of light by these three cell types, we can more or less detect every hue in the visible spectrum. Many animals only have two photo-receptors. Some, including birds, have a fourth that absorbs ultraviolet light. Mantis shrimps have *twelve* different colour receptor types, plus those which can discriminate both linear and circular polarised light. This is unheard of elsewhere in the animal kingdom.

If three or four types of receptor is sufficient to cover the known range of visible light, why mantis shrimps have so many has long been a puzzle. It was assumed that they saw the world in a glorious rainbow inconceiv-able to the human brain – and indeed this theory seemed to suit their incredible array of vibrant colouring (sadly when crustaceans are pre-served in fluid in museums all of the pigment leaches out). They clearly use flashes of colour in communication with each other.

However, it turns out that they do not see like any other known animal. Rather than compare the signals between neighbouring receptors on the light spectrum, like other animals, mantis shrimp simply recognise the twelve colours that their receptors can detect. Instead of having com-plex computational comparisons of colour responses between different receptors, their cells are simply firing when they detect their one specific colour. It is assumed that this allows them to process visual information far quicker, and without much brainpower – clever adaptations to living in the high-speed world of ambush predators. It allows for minute adjust-ments to their rapid attack weapons after they are 'fired', if the prey moves.

Colour perception aside, the eyes are also arranged in ways that help a predatory lifestyle dependent on accuracy. Their eyes are on stalks that are independently mobile, constantly scanning the environment for food. Each eye is then split into three zones in a way that gives each zone a slightly different perspective on a similar field of view. This means they only need the image from a single eye to achieve depth perception (we need to compare the images from our two eyes to judge distance). This is another adaptation to their incredible predatory behaviour.

70 Surinam Toad

SPECIMEN PRESERVED GIVING BIRTH

Parental Care

In the great game of evolution, an animal's key mission is to ensure its genes are passed on to future generations, for as many generations as possible. From a parent's point of view, having a baby is not enough – it is critical that offspring live long enough to breed for themselves, passing on the parents' genes to their grandchildren.

This means that it is in an individual's interest – evolutionarily speaking – to ensure that they have babies that survive. Animals normally tackle this with one of two tactics: stack the odds by having a huge number of youngsters and hope that a few will survive just by sheer weight of numbers; or produce fewer offspring but give them as much help as possible by providing a lot of parental care.

Neither approach is particularly 'better' than the other – both require a serious energetic investment at some stage (at least for one sex). Producing huge numbers of fertilised eggs or tiny babies is energetically expensive for the female, while caring for developing young comes at a cost for whomever is left holding the baby.

European common frogs are typical examples of the reproduction-by-numbers strategy. Although there are risks that their eggs will die due to predation, fungal infection or environmental conditions, they lay so many eggs in a clump of frogspawn that chances are that a good number will survive to emerge as tadpoles, and at least a few will make it through metamorphosis to become adult frogs.

Mammals, on the other hand, are extreme parental carers – we produce a small number of babies at a time, and invest significantly in their development: at first providing breastmilk and then helping them to find food until they are independent enough to feed themselves. This can take years of investment from the parents.

What's interesting is that both tactics are often seen within the same taxonomic group. For example many fishes produce hundreds of tiny babies which have to fend for themselves, while others dedicate themselves to raising their young. As discussed elsewhere in this book (chapter 35) male seahorses carry their eggs in a pouch until they are ready to hatch. A number of cichlid fishes care for their growing young in their own mouths.

Amphibians are also diverse in their parenting strategies. Aside from the mass spawnings of many frogs and toads, there are a number of species that use different methods to care for a smaller number of eggs to increase the chances of survival. For example two species of Australian

AMPHIBIA 3.1
Anura
Opisthocoela
Pipa americana

gastric-brooding frogs swallowed their young so they could develop in the safety of their own stomachs (sadly they became extinct in the 1980s); male midwife toads carry a small number of eggs tied in a string around their legs; and marsupial frogs carry their developing eggs in a pouch.

The Surinam toad has also evolved a bizarre method of being a good mother. This species lives in the Amazon where there are plenty of predators to avoid – from piranhas and caimans to river otters and dolphins. This helps to explain why adult Surinam toads are shaped rather like a rotting leaf, and they have the drab muddy brown colour to match.

Their bodies are extremely flat and while their legs are strong for swimming, their arms are held out in front of them to maintain the flat shape, rather than held under their bodies like other toads and frogs. They rarely venture on to land, which is a good thing as this position stops them from being able to lift their bodies up with their arms. They have a wide, triangular head and little flanges on their sides to help them blend in with the muddy river bottom. Such camouflage is great for the adults, but wouldn't suit a tadpole. That's fine though, as Surinam toad tadpoles never swim free.

When mating, the males and females perform the same embrace as other frogs and toads (called amplexus), but as the female lays the eggs they flip over so they are belly-up and the male is below the female. He catches the eggs, fertilises them and pushes them onto the female's back, where they stick. After their encounter the female sits motionless for a few days so not to dislodge the eggs. The skin on her back begins to swell and grow around the eggs, eventually engulfing them completely in little skin pockets.

The female then goes on living her camouflaged, muddy life for three or four months, during which time the eggs develop into tadpoles, and then into mini toadlets, all in the safety of the covered pockets on her back. When the time comes, the tiny toads then burst through her skin in a manoeuvre reminiscent of the film *Alien*, ready to face the world as a small camouflaged toadlet.

What's slightly unpleasant/interesting about this specimen is that it has been prepared mid-'birthing'. This may seem rather cruel, but to make the specimen as instructive as possible, and show some closed pockets and some emerging young, it's clear that some of the toadlets were allowed to break out before the animals were killed. In their preparation, they were then unceremoniously pushed back in – we can see this as many of them appear to be coming out backwards, which they would not do in the wild.

Surinam toads are also unusual in their feeding – they do not have the tongues so typical of their relatives. Instead, they shovel small fishes and invertebrates such as worms into their mouths with their sweeping arms. To detect food in the murk, they have incredible star-shaped projections from each of their fingertips which sense their prey.

71 Olm

Animals That Never Grow Up: Peter Panimals

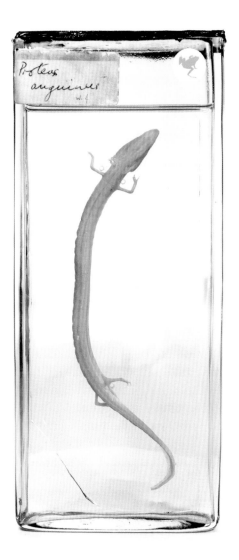

The vast majority of animal species have a larval form and an adult form (and sometimes more than one stage in between). To transition from one to the other they have to go through metamorphosis – the spectacular rearrangement of their body, with the death and reabsorption of their larval features, and the appearance and growth of their adult ones. The definition of an adult is an individual that can reproduce.

There are various evolutionary reasons why separate larval and adult stages exist. Often eggs cannot be produced with a big enough yolk to allow the embryo to complete its development before hatching. A simpler larval stage therefore allows the animal to feed itself in the environment before metamorphosing. Larvae and adults also tend to have different diets (most frog tadpoles are herbivores, while adults are carnivores, for

Adult olms – a kind of cave salamander – never transform from their juvenile form.

227

example), or even live in different habitats; meaning that the parents are not competing with their own young for resources.

Frogs and insects are most famous for their metamorphoses, but actually it's seen in nearly every major phylum of animals. Among the vertebrates, most bony fish have larval stages, as well as the amphibians. Perhaps the most dramatic is the rearrangement of the flatfish from their larvae. When they hatch, juvenile flatfish are symmetrical and look much like other fish larvae. However to become the flattened adult, many of the elements of the skull are completely rearranged, and one eye migrates across the face so that both eyes point upwards (there's no point having an eye that faces into the mud), as well as reorganisation of the fins.

However, some animals never truly grow up, and become adults whilst retaining at least some of their larval characteristics – this process is called paedomorphosis. One of the most extreme examples is the olm –

a cave salamander from the mountains of Slovenia and neighbouring countries. Like its North American cousins the mudpuppies, olms never metamorphose from their juvenile form. Living deep in aquatic cave systems, olms are incredibly well adapted to an environment with very little food. As such, their life is not very fast paced. They take 7–12 years to sexually mature, and can live for nearly 60 years. They have been shown to be able to survive for 10 years without food.

In their cave pools, there would be no advantage to acquiring the terrestrial mode of most adult salamanders, and so they forego

Although technically an adult, this axolotl has retained its larval features like external gills.

metamorphosis. Larval amphibians have external gills: fluffy fans of blood-rich tissue that allow them to absorb oxygen readily from the water. These are lost during metamorphosis. By keeping hold of these juvenile features, adult olms can stay in their aquatic environments for their whole lives.

Another salamander is also famous for paedomorphosis – the axolotl. Although it has been widely used as an experimental lab animal (it has amazing abilities to regenerate limbs) and as a pet, axolotls may be extinct in the wild in the mountain lakes of their native Mexico. Paedomorphosis is regularly seen in salamander species that live in high altitudes, as the land around the aquatic habitats where the juveniles live (where gills come in handy) can get very cold or lack regular food. Unlike olms, axolotls can metamorphose if the conditions are right.

An incredible example of paedomorphosis in insects is seen in the Strepsiptera (twisted-wing parasites) which live inside other insects, including bees, flies, bugs, grasshoppers and cockroaches. Adult females live their entire lives buried within their hosts, with an opening to the outside world through which males (which only live for a few hours) hypodermically inject sperm, and the young larvae escape. These legged larvae then find their own host, burrow in, and go through a series of moults without ever actually leaving their older exoskeletons, creating a protective layer of un-shed casings around themselves, and finally a pupa. The metamorphosed males then leave their hosts to find a buried mate, while the paedomorphic females, which are legless and wingless stay put.

Paedomorphosis has been proposed as an important evolutionary mechanism. The retention of larval features and larval ways of life by individuals which develop mature sexual organs could cause the evolutionary separation of those that metamorphose and those that don't, creating opportunities for new taxonomic groups to arise. Indeed there are entire groups of fossil amphibians that appear to have arisen as paedomorphic versions of their ancestors. Modern lungfish also look as though they came about through paedomorphosis, retaining larval features seen in the bony fish they evolved from.

For a long time, paedomorphosis was even believed to explain the origin of vertebrates. As discussed above (chapter 18) the sea squirts are non-vertebrate members of our phylum, the chordates. This is evidenced by the fact that they have a tadpole-like larval stage with a notochord – the precursor to the spine. It was suggested that vertebrates evolved from a sea-squirt larva that retained its juvenile features but developed reproductive organs. As tempting as this theory is, it is contested by recent molecular studies that suggest that sea squirts followed a different evolutionary path to vertebrates.

72 Leaf-Cutter Ant

PINNED SOLDIER

Family Relations: Haplodiploidy

Family relationships can on occasion be less than straightforward (at least among humans): there can be feuds, sibling rivalries and accusations of favouritism. Therefore, it might be simpler to consider them on a purely mathematical basis. It is possible to calculate the extent to which two individuals are related to each other, as a measure of the statistical odds that they will share any given gene. This can be interesting as we can reap evolutionary benefits from our relatives if they breed.

If we consider the crux of natural selection to be about maximising the opportunities to pass on your genes to the next generation, it is important to remember that having babies yourself is not the only pathway to success. As your relatives share some of your genes, you will also 'win' if they reproduce, and pass on those genes too. The more closely related you are to someone, the more evolutionary benefit you gain when they breed.

In most sexually reproducing animals, individuals get half of their genes from their mother and half from their father. This comes about because sex cells – eggs and sperm – only contain half the amount of genetic information as body cells. Humans, for example, have forty-six chromosomes in twenty-three pairs. However our sex cells only contain twenty-three single chromosomes (corresponding to one from each pair, but actually segments of the DNA get swapped over between the two halves of each pair when the sex cells divide).

Therefore when a sperm meets an egg, a new individual is created with a new combination of forty-six chromosomes. This means that you share half your genes with your mother and half with your father. You are related to them by a half, and they are related to you by a half.

Likewise, if you have the same parents you are related to your brothers and sisters by a half. The genes that your mother passed to you will not be the same as the genes that she passed to your sibling – it is not an identical 50 per cent. Because only half her genes are passed on in any one egg cell, the odds that you share a gene from your mother with your sibling is 25 per cent – half of a half. The same applies to the genes from your father, so adding them together, you are related to your siblings by a half (you share a quarter of your mother's genes and a quarter of your father's).

When your siblings breed, they will pass on a half of your shared genes to your nieces and nephews, which means that it makes evolutionary sense for you to help your parents and siblings to raise their young, at least to some extent.

Such cooperative breeding is seen to an extraordinary degree in some insects. Bees, wasps, ants, termites and aphids have some astonishing breeding systems where many members of a colony dedicate their entire lives to raising someone else's children – many of them are even sterile. Caste systems arise with different individuals having different roles, all helping one single queen to reproduce.

Leaf-cutter ants, like the one pictured opposite, are fascinating animals that develop incredibly complex societies, with differently sized and shaped ants performing different roles, all to the benefit of the queen. Living up to their name, leaf-cutters bite off chunks of leaves and carry them back to the nest, travelling in great columns. The largest ants – like this soldier or major worker – have huge heads with strong jaws to defend the column (and the nest) from intruders, as well as cutting through thicker parts of plants and clearing obstacles from the path. Smaller ants carry pieces back to the nest.

However, leaf-cutter ants don't eat the leaves: they use them to farm a fungus which grows in the warm, humid protected environment of the ant nest. There are castes dedicated to tending the fungus garden, feeding their crop to the developing larvae and carrying out waste. All for the benefit of the queen and her young.

In other social insects, castes have developed some incredible roles and physical attributes. There is a termite worker with a head shaped like a bung. Their role is to be a living door, using their heads to plug the nest opening. Some honeypot ants act as living larders – swallowing so much food that their abdomens swell into grape-like balls. They then hang from the roof of the nest (unable to walk anywhere due to their girth). Periodically other workers will come and extract a meal from them. Despite their widely different anatomies, different castes within a colony

Different termite forms: a nymph, a queen, winged females and a wingless male.

do not differ genetically – their fate depends on the conditions and diet from when they were raised as larvae.

A key theory for how such selfless behaviour evolved in bees, ants and wasps is their unique genetic system, called haplodiploidy. Females are produced through sex – gaining genes from their mother and their father (like us they have chromosomes in pairs – they are diploid). However males are produced asexually from unfertilised eggs (they have no father), which means their entire genetic make-up is half of their mother's genes, in a single set of chromosomes (they are haploid).

As males are haploid, they do not divide their genetic information to make sperm, so they pass on 100 per cent of their genes to their daughters (and males can never have sons). This means that (assuming their mother was monogamous, which in many species they are) sisters share all of their father's genes with each other, and – as in humans – a quarter of their mother's. In total, therefore sisters are more closely related to each other (75 per cent) than they are to their own young (50 per cent). So sisters will benefit more by helping their mother to raise more sisters than by having offspring themselves. It is believed that this is one factor that allowed for such an extreme form of apparent selflessness to evolve – being sterile for the benefit of your colony.

73 Aphids

PRESERVED SPECIMENS

Virgin Births

When I was young I had a pet stick insect. Despite the fact that I'd had her since she was a baby (technically, a nymph), and she lived her entire life *alone* in a jar, once she had grown big she started to produce eggs. Lots and lots of eggs. Obviously this isn't remarkable, as animals produce eggs without mating all the time – this is exactly what the hens' eggs are in the supermarket, not to mention the human menstrual cycle.

However, what was surprising (at least to my 10-year-old self) was that many of these eggs would actually hatch, with tiny perfect stick insects hauling themselves out of the little lozenge-shaped capsules. How could this happen? Luckily my mother was a biology teacher, and so I learned early about the process of parthenogenesis.

Parthenogenesis, from the Greek for 'virgin creation', is the process by which embryos develop without being fertilised – they are produced without their mother having sex. We've already discussed this with the haplodiploid ants in the last chapter – male ants, bees and wasps develop out of unfertilised eggs, with a single set of unpaired chromosomes. This is one example of parthenogenesis, but there are many more which involve individuals developing with two sets of chromosomes, both of which come from their mother.

Actually, parthenogenesis occurs in all major animal groups except one: mammals are the only group where asexual reproduction of adults is unknown.

Parthenogenetic youngsters can be produced in a few different ways, the simplest being full cloning: female aphids, for example are full clones of their mothers. They are produced by an egg being formed with the exact paired set of chromosomes of the mother – they just divide a cell in two to create an egg, but without dividing the chromosomes as they do in sexual species. Male aphids are similarly developed, but they only receive a single sex chromosome from their mother – males are XO and females are XX (receiving both their mother's sex chromosomes).

More genetically complex systems create individuals which are half clones of their mother. Although they receive all their genes from their mother, they are not genetically identical to her as they don't receive both copies of all her genes. This can either occur by the production of an egg cell with half the mother's genetic information in it, which then gets replicated (producing two copies of each chromosome); or by the fusion of two divided pairs of chromosomes – the latter could technically be considered sexual reproduction as new combinations of genes are being formed.

If females can produce babies without going to the trouble of mating, why do males still exist? After all, sex is an energetically expensive and often violent activity. Not to mention the fact that the mere existence of males – which can't give birth – uses up a huge amount of resources (food, shelter, water etc.) that might be more profitably directed to the females. Isn't it a waste if half of your population can't even produce young? Are males anything more than parasites?

This question has long puzzled biologists, particularly as completely abandoning sex is rather rare in the animal kingdom. Most species that completely give up sex tend not to survive for very long, although there are a few exceptions – bdelloid rotifers, for example, are microscopic invertebrates that live on moss, and they have gone without sex for tens of millions of years.

Some examples of asexual reproduction have been found among captive animals in zoos and aquaria, with keepers being surprised to find babies even though there is only one sex in the enclosure. This has been particularly well documented among lizards and sharks. However, because the rate of hatching of the eggs is so low it is believed that these are examples of accidental parthenogenesis rather than being an adaptive response to a life without males. Something has gone wrong with the production of unfertilised eggs, and they've ended up with both halves of a chromosome pair.

A female aphid giving birth to a clone. The youngster itself may already be pregnant with another clone.

There are significant evolutionary advantages conveyed by sex, specifically from the genes being shuffled. This helps sexual animals to avoid accumulating too many harmful mutations (sex weeds these out), and creates genetic diversity which makes it difficult for parasites and diseases to take hold. Sex produces a moving target – without genetic mixing parasites only have to find a means of beating one immune system.

However, asexual reproduction through parthenogenesis can be a handy tool for species to make use of on occasion. Alternating between sexual reproduction and parthenogenesis could be triggered by environmental conditions that make it fruitful to suddenly increase in number: because of the hassle involved in sex, and the fact that males keep eating your food, asexual reproduction is usually quicker.

Some aphids, for example, all hatch out of eggs as females in the spring. They then reproduce quickly parthenogenetically to make immediate use of the season's bountiful food by giving birth to live female clones of themselves, who rapidly produce many more generations (in fact some are born with another clone already developing inside them, like a Russian doll).

Towards the autumn they begin to give birth to males (who are also genetic copies of their mothers, but lacking one of the sex chromosomes). Males and females then mate sexually, and the female produces eggs which will overwinter, ready to start the cycle again.

74 Eyed Hawkmoth x Poplar Hawkmoth Hybrid

Hybrids

It may be a little late in this book to remind you that taxonomy is entirely made up. Humans have put animals into groups based on differences they perceive between them. These may be easily obvious differences, or

A hybrid moth – its parents belonged to two different species.

they may only be observable under the microscope (insect species particularly, are defined by the minute differences of their genitalia) or even at the genetic level. Some species are even split on the basis of their song.

Unfortunately there isn't a specific *amount* of difference that two individuals or populations have to be in order to be classed as different species. This makes things rather subjective. For this reason, the most widely used definition of a species is a group of individuals that are similar enough to each other that they can interbreed and, crucially, *produce fertile offspring*. Plenty of things can mate together, and some can even produce young, but if that young can't grow up to have babies of its own, its parents may not have belonged to the same species. This is known as the biological species concept. It's the best we've got, but it doesn't work particularly well. It's all made up.

We have to think of at least three layers of mechanisms that get in the way of two individuals producing fertile young together: having sex, viability and being fertile. First of all there are mechanical considerations of having sex: can two individuals actually have sex? To have sex you have to be in the same place at the same time, and be reproductively receptive. Many insects, for example, avoid accidental pairings with closely related species by emerging from their pupae at different times, so they physically avoid each other. Other animals only release their eggs ('come into heat') at a certain time of year, and that varies between species.

There is also the actual mechanical act of sex – does the male's penis fit inside the female's vagina, does the female have the physical attributes to receive the male's sperm?

Then, if they get past that hurdle can a mating produce viable offspring? Can the sperm fertilise the egg? Before it gets a chance the female's immune system may recognise that the sperm in her system is 'too different', and kill it. But if it does reach the egg, and inject its genetic information into it, can it develop into an embryo, and eventually be born?

One of the key considerations here is that the genetic make-up of the mother and father may be so different that they can't pair up their chromosomes, or that the new combination of genetic information doesn't have compatible instructions to actually grow the foetus. Matings between members of different species often fail at this stage.

The final test of the biological species concept is if an embryo does go to term, gets born, and grows to be an adult, can it produce young of its own: is it fertile? Often two closely related species have chromosomes in a different number of pairs, so when they come to combine during fertilisation, they can't all match to their opposite number from their mate. This may not prevent the animal from growing, but when they come to divide their paired chromosomes to create sperm or eggs of their own, the division is imperfect, and the sex cells aren't fertile.

So, if two individuals can get past all these ecological, mechanical, genetic and developmental barriers and produce fertile offspring, then they can be considered the same species.

Unfortunately, there are plenty of examples of this not being the case, and two individuals that have been placed in different species by taxonomists do manage to mate and have grandchildren. The insect pictured on p. 236 is a hybrid between two different species: the eyed hawkmoth and the poplar hawkmoth. This is particularly remarkable, as taxonomists hadn't even put these two species in the same *genus* (the taxonomic group above the level of species), so they didn't even think they were particularly closely related. The resulting hybrids can produce fertile young, but it is often weak and can't produce young of its own (the original pair can have grandchildren, but not necessarily great-grandchildren).

Eurasian humans (the people of Europe and Asia) today appear to be the result of hybridisation in our distant past. Once the Neanderthal genome was sequenced it could be compared with modern humans. Eurasians were found to contain 1–2 per cent of Neanderthal genes. The most likely explanation for this is that our ancestors hybridised with Neanderthals (which may explain my prominent brow ridge).

The human evolutionary tree is complicated and our understanding of it changes regularly, but our species, *Homo sapiens* is a descendant of *Homo erectus* that evolved in Africa. Conversely Neanderthals, *Homo neanderthalensis* are descendants of *Homo erectus* that evolved in Eurasia (*Homo erectus* was found in both places). When the ancestors of Eurasians left Africa, it appears that they bred with the Neanderthals they met, and our genomes today reflect that. This means that the hybrids were fertile. Rather than a forking family tree, it seems that on occasion branches can fuse back together.

Although it is extremely convenient for us to consider species as discrete units, separate from one another, nature in reality is much more fluid. Hybrids are a reminder of that.

75 *Archaeopteryx*

CAST OF NEAR-COMPLETE FOSSIL

Winging It: The Evolution Of Flight

This is a cast of one of the most famous fossils in the world, evidencing the link between birds and the dinosaurs from which they evolved. The original is in London's Natural History Museum, having been purchased in 1862 from a German physician called Karl Häberlein soon after its discovery. That is called 'the London specimen', and it is one of just eleven fossils of the skeleton of *Archaeopteryx* to have been found. They all come from the incredibly fine lithographic limestone quarries near the Bavarian town of Solnhofen in Germany.

At about 146 million years old, *Archaeopteryx* is considered by most to be the oldest known bird (others periodically claim that it is not quite a bird, and is instead simply a non-avian feathered dinosaur, just outside the birdy lineage). As such it has generated a lot of interest in the last century and a half. The Solnhofen sediment has preserved the impressions of its wings, and most of its skeleton. Unlike modern birds it had teeth and a feathered tail supported by its spine. But like both birds and non-avian dromaeosaurid dinosaurs, it has a wishbone or furcula to support a strong shoulder joint, and a special rotatory joint in its wrist. In dromaeosaurs like *Deinonychus* these features enabled them to catch prey with an in-swinging motion of the arm: a pre-adaptation that enabled their bird relatives to take flight.

Archaeopteryx's feathers were asymmetrical, unlike those which have subsequently been discovered on the non-avian dinosaurs from which it evolved. Asymmetrical feathers create an aerofoil like a plane wing, generating the lift necessary for flight. From this we can see that *Archaeopteryx* could certainly fly, but not as ably as modern species as it doesn't have a large keel on its breastbone to support huge wing muscles.

When it was first discovered, some anatomists were quick to point out that it had this mix of both reptilian and bird-like features. This was exactly the kind of thing that Darwin had proposed would be found to support his theory of natural selection, published just 3 years earlier. One of Darwin's most prominent critics was Richard Owen, one of the world's leading scientific authorities, and Owen wanted to get his hands on it. He was the superintendent of natural history at the British Museum (which would split to form the Natural History Museum), and forked out £700 for Häberlein's Solnhofen collection.

Owen concluded that it was merely a long-tailed bird fossil (failing to notice/mention elements of the skull that would prove otherwise), and dismissed the notion that it provided evidence for evolution. When Darwin's most vocal supporter, Thomas Henry Huxley got to see it, he pronounced that it was a bird that closely resembled dinosaurs. This was the first serious evidence supporting what we now know – that birds are highly adapted dinosaurs; and the only dinosaur group to survive the mass extinction at the end of the Cretaceous.

The specimen from the Grant Museum of Zoology pictured on p. 239 may be a cast, but it should not be written off as unimportant as it has real historical significance: it was one of the first casts ever made, soon after *Archaeopteryx*'s discovery. Subsequently, the original has been prepared further with some of the surrounding limestone having been removed. This gives us a snapshot of the evidence that was available to scientists in the past – they would have seen a different specimen to what is at the Natural History Museum now.

Birds of course were not the first group of animals to take to the air, nor were they the last. Insects beat them to it by around 200 million years. Flying has obvious advantages for a vertebrate if the mechanical challenges to achieve it can be overcome. With so many flying insects to eat, the power of flight opened up a whole new hunting opportunity. It also allowed for immediate escape from predators; and freed up the feet so they could be used in feeding rather than walking.

Despite these drivers, powered flight has only evolved three times in the vertebrates: in pterosaurs, birds and bats. Pterosaurs were the first, flying on a wing formed of a single finger (the fourth), from the Late Triassic period, around 50 million years before birds. From this extraordinarily long digit their wing membranes stretched down, stiffened by fibrous rods.

Pterosaurs likely filled many of the ecological niches that birds do today. Pterosaur fossils had skulls and teeth adapted to catching fish, insects and small vertebrates (including birds), crushing hard foods like shells or tough fruit, and even sieving plankton out of the water. Birds and pterosaurs both have long and diverse fossil records, so they were/ are both highly successful. After birds appeared, the smaller pterosaurs tended to decline throughout the Jurassic and Cretaceous, but truly giant ones evolved: *Quetzalcoatlus* had an astonishing wingspan of 10–11m. Pterosaurs were the largest fliers ever.

Bats were the last vertebrate group to take to the air, during the Eocene period a little over 50 million years ago. Their wing is different again, formed of a membrane of skin stretched between four elongated fingers: they flap with their whole hands, using their tiny thumbs for climbing.

A fossil of the Jurassic pterosaur *Rhamphorhynchus muensteri*. Although the wing bones are not visible in this specimen, the stiffening rods that supported the wing flaps have been preserved around the tail.

76 House Mice

COLLECTION OF 6,000 SKELETONS

Island Evolution And Ecology

One wall of one of the store rooms in the Grant Museum of Zoology at University College London is dominated by a collection of around 6,000 mice skeletons, each contained in their own glass vial, grouped in boxes depending on the place in which they were caught. It is, by far, the largest collection of a single species in the museum. Why would a relatively small museum have such a huge collection of one species? And why such an everyday species as the house mouse?

The mice come from locations all around the world, and very many of them come from islands. During the 1960s and 1970s a zoologist working at University College London called R.J. Berry caught these mice (many by hand) in order to use them to investigate some of the interesting things that evolution and ecology do to species on islands.

House mice are a great choice for studying these processes as they are a single species found all around the globe. The species is native to northern India, but found great success with living around humans, and they spread widely following the advent of agriculture. Following that, humans have more recently transported them all across the planet as stowaways.

Among the evolutionary processes that R.J. Berry used the mice as evidence for was the 'founder effect'. When islands are colonised by a small group of individuals of a species from a nearby mainland, the colonising 'founders' of the new population will not have the same genetic diversity as the parent population. This can make the islanders appear very different from the average mainlander.

In humans, for example, we see a wide variety of hair colours. If we were to randomly select fifteen people to start a settlement on Mars, there is a chance that among them not all the hair colours are represented. This could mean that the population they build on Mars doesn't have any red-heads. It's also possible that all of the founders have the same hair colour, which means the new Martian colony could start without the genes for any of the other varieties seen on Earth.

The founder effect on islands results in a population that has much less genetic variety – a shallower gene pool – in which some genes or traits are not seen at all, and some are seen in much higher proportions than in the population they came from. This is also discussed in the section on dispersal (chapter 44).

R.J. Berry also used these mice to investigate the effect of climate on body size within a species – a theory known as Bergmann's rule. This states that, for a species that has a wide geographic range, individuals that live in colder environments will grow to larger sizes, and those that live in warmer places will be smaller. This is because larger animals (or indeed objects) can retain heat more efficiently – they have less surface area relative to their volume, and so less skin through which heat can escape.

Mice are perfect for this study, because they are found all over the globe and so comparisons can be made between populations in different climates. Islands add in an extra factor as they are typically harsher environments to their nearest mainland. Using mice from islands around the coast of the UK as well as the sub-Antarctic, Berry found that body size did indeed increase with decreasing temperatures.

Incidentally, to remove the error created by mistaking a young mouse for a small mouse, Berry had to come up with a method for accurately ageing the skeletons. The extent to which teeth are worn down is a good way to approximate age – looking at how much enamel has been lost from a tooth, exposing the dentine below. Berry found that if he dipped the mice skulls in hot tea, the more porous dentine would become stained, but the enamel would not, allowing him to work out the ratio between the two.

Islands can also do amazing things to the size of an animal for other reasons – there are plenty of examples where small animals have evolved to become giants on islands, and large animals become dwarves. This will be explored more fully in the next chapter, but it was found to some extent in the house mice. As islands can have fewer predators (they may not be big enough to sustain them), the mice in cold climates could grow even bigger as they lost the need to squeeze into tight crevices to hide. In this case Bergmann's rule was just one factor in the evolution of body size.

77 Elephant Bird

CAST OF EGG

Island Giants And Dwarves

Within very recent history – during the seventeenth century – there were some colossal birds still walking the earth. Walking, but not flying. The elephant birds of Madagascar could potentially reach 450kg in weight – that's three times the size of the largest ostrich, the avian record holder today – and around 3m tall. They were among the largest birds to have ever lived. Needless to say that at that size they had lost the power of flight.

This is a cast of one of their eggs, the remains of which have been found across Madagascar along with sub-fossil skeletons of the birds themselves. The majority of specimens are shards of broken shell, but complete ones are occasionally uncovered and some even have the remains of developing chicks inside. In fact, some have been found in the sand dunes of Western Australia, thousands of miles across the Indian Ocean, having floated there on ocean currents.

Elephant bird eggs have the same volume of 160 hens' eggs. This one is over 30cm long. It seems reasonable to assume that they were collected by humans once they arrived on Madagascar a couple of thousand years ago; they could have fed a sizeable party. Although humans don't seem to have hunted the adult birds a great deal, egg poaching (excuse the pun), habitat loss and climate change are likely to have combined to seal their eventual extinction in the 1600s.

They belong to the group of birds called the ratites, which are all flightless except the South American tinamous: the kiwis and moas of New Zealand; the emus and cassowaries of Australasia; the rheas of South America; the ostriches of Africa and the elephant birds of Madagascar. Despite Madagascar's geographical proximity to African ostriches, a 2014 study looked at the DNA of museum specimens of this species and found a surprising result: their closest relatives are the kiwis of New Zealand.

This is remarkable as it had been assumed that the flightless ratites had achieved their wide distribution on the southern continents through vicariance (see chapter 45): that they had been present on the super-continent of Gondwana before it split up into the modern continents. The DNA results means that the ancestor of one or both of the kiwis and elephant birds must have *flown* to their current homes, as the evolutionary split occurred after the continents separated.

What happened to elephant birds then is really interesting: they grew to be giants.

There is a fascinating trend that operates on islands across the world: small species tend to get bigger and big species tend to get smaller. It's called the 'island rule'. The evolutionary drivers that act on body size are very different on islands than they are on the mainland, as often islands contain far fewer species, and fewer things to eat.

On the mainland, small species have to contend with avoiding predators and competing with other species for food and shelter. This can be achieved by being small enough to escape into tiny burrows, and being small enough to live alongside competing species: large bodies need more food, which is difficult to find if several species are after the same resource.

Islands tend to have lower species diversity (a trend which grows stronger with smaller and more remote islands), but higher densities of the species that do live there – there are more individuals of a given species in a given area. They tend to have few large predators, so evolutionary pressures to avoid them drop away. Animals on islands tend to be competing for resources with their own kind, rather than other species. The best strategy for that is to grow bigger and outmuscle them. Both of these factors drive smaller species to grow giant on islands.

There are examples from across the world where species have grown giant after reaching islands – including many species of giant rodents and shrews – even the giant lemurs of Madagascar (see chapter 48) are an example of this. Dodos, moas, kakapos and many other birds follow the same trend. And with birds there is also the question of flight: without predators, escape by flying becomes unnecessary so their wings shrink. And without the need to fly, the pressure to remain lightweight is gone.

The founder effect could also play a role here (see the previous chapter), as larger individuals are more likely to survive the sea crossing from the mainland than smaller ones, and so the genetic make-up of a newly arrived species is already different from those it left behind across the ocean.

The opposite trend is seen for animals that were ancestrally large. For animals like cattle, deer, mammoths and elephants, being large is a defence against predators – they are simply too big to be hunted. But this comes at a cost, as you have to eat more if you are big. On islands, where the predators are rare the costs of size outweigh the benefits: species are released from the need to outgrow predators.

Island dwarves are seen in the fossil record of the Mediterranean islands, where dwarf mammoths, elephants, cattle, deer, goats and hippos have been found. The trend is repeated across the world, and may even account for the miniature hominid species *Homo floresiensis* found in Indonesia in 2003.

Unfortunately, both island dwarfism and gigantism result from a lack of pressure from predators, and this left animals at both ends of the scale highly vulnerable to extinction when humans arrived, and very many examples of both are now only known from their remains.

Giant elephant birds survived on Madagascar until a few centuries ago.

78 Thylacine

SKULL

Extinction And Over-Hunting

One of the most important evolutionary processes is extinction. Pretty much every species that has ever lived is now extinct. Extinctions come about through competition, when new species arrive or evolve that are more successful than the previous ones; and catastrophe, when events suddenly stop a species from being able to survive in its habitat. These can be local or global: floods, fire or disease could ravage a species' entire range, for example; or the world could be altered by meteor impact or dramatic climate change. One such catastrophe is the evolution of humans.

The following chapters will explore the topic of extinction from a number of themes. Species become extinct all the time – it is a natural process – but never before has a single species wrought such destruction on such a scale. One way in which we have impacted biodiversity is through over-hunting. Unfortunately, I could choose from thousands of species to tell this story, but I'm going for one that is extremely close to my heart: the extinction of the thylacine.

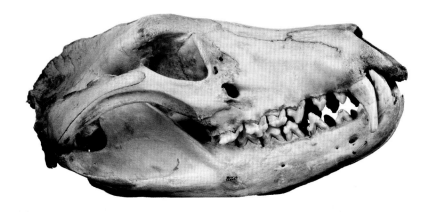

Thylacines at the US National Zoological Park, Washington DC, photographed in around 1905.

Picture this: an animal in a zoo dies of exposure because the door allowing it to return to the inside area of its enclosure was accidentally locked shut overnight. It is early spring and southern Tasmania gets pretty cold – a wire and concrete cage is no place for a warm-blooded creature to be kept outside.

This is what happened to the last known thylacine, on 7 September 1936. The neglect itself would be shocking for any animal, let alone the sole known member of a species, but especially when that species was the only one in its entire family. That day, a whole branch of the tree of life fell off. Well, in truth it was cut off.

Thylacines were the largest then living marsupial carnivores – related to kangaroos and koalas, but shaped very much like a dog with stripes (another example of convergent evolution). They were also called Tasmanian tigers. They once ranged across New Guinea and mainland Australia, but disappeared there around 2,000 years ago. Tasmania was their last refuge.

Thylacines are now on the logo of the island state (and one of its most popular beers), but this shows a significant change from the status with which they were held in the century prior to their extinction. Newly settled Tasmania was attempting to build a sheep farming industry, but suffered heavily with loss of livestock. At the time the powerful farming lobby blamed the thylacine, and successfully campaigned for a bounty to be placed on its head. The eventual extirpation that resulted from this government-endorsed slaughter is the only example of a deliberate extinction that I can think of.

That would be bad even if thylacines *did* have an impact on sheep-farming, but we now know that the evidence shows that they very rarely hunted sheep. In truth it was dogs that were allowed to run feral, and it was humans themselves stealing the livestock.

A number of private bounties had been in place for killing thylacines since 1830 (which is pretty remarkable considering the first (small) European settlement was only established in Tasmania in 1803), and following the presentation of some decidedly questionable facts at parliament, the government sponsored a state-wide bounty in 1886. This account written 8 years later is disquietingly ominous considering what we know now:

> The Thylacine was at one time an abundant animal in its native island. The damage which it inflicts on the flocks of the settlers has, however, given rise to a relentless war of extermination, which has resulted in the almost complete extinction of this, the largest of the Australasian Carnivores, in the more settled portions of the country.*

As I say, the contention that thylacines significantly impacted the sheep industry is now believed to be inaccurate. Numbers continued to fall, but, because of pressure from the farmers, nothing was done. Eventually, the Tasmanian government did place the thylacine under protective legislation, in 1936. It didn't work – the last known individual died that very year.

Museums are now the only habitat for the thylacine, and for me they are the greatest icon of extinction. Dinosaurs and mammoths are unquestionably awesome, but their extinction is, to me, far less meaningful than the thylacine's. Humans did most likely play a role in the mammoths' demise (though obviously not the dinosaurs'), but I find the concept of both to be relatively abstract. I know they were real (we have their remains in museums), but in a sense, at least to me, they may as well have never existed. The idea of them is incredible, but mammoths and dinosaurs only really exist in our museums, in our imaginations and on our cinema screens.

I spend a lot of time in Tasmania on fieldwork – it is one the most incredible places for wildlife on earth. While I do not believe (as many do) that thylacines still live there, hidden in the vast wilderness, when I sit and look across many of Tasmania's habitats I constantly imagine them there, in a way that I don't imagine dinosaurs and mammoths when I visit the moors of Britain. I think the reason is that thylacines are part of our very modern history, in living memory, and I sense the blood on our hands.

---

*   Lydekker, R., *A Hand-book to the Marsupialia and Monotremata* (E. Lloyd: London, 1894)

79 Tasmanian Devil

MOUNTED SKELETON

Extinction And Wildlife Disease: Contagious Cancer

The thylacine in was the world's largest marsupial carnivore. When it disappeared, that mantle passed to the Tasmanian devil. Tasmanian devils are about the size of an extremely muscular badger, or an English bull terrier. Tasmania is absolutely full of food if you are a Tasmanian devil: kangaroos, wallabies, pademelons (like a wallaby, but smaller), wombats and possums are everywhere. Devils can hunt, but they are most famous for their scavenging adaptations, with huge hyena-like jaws for crushing bone. They can eat every part of a carcass and have an impressive set of teeth. I once saw one climbing out of the anus of a dead pademelon, while two others were working their way in through the back. Everything was going pretty well for the devils, until 1996.

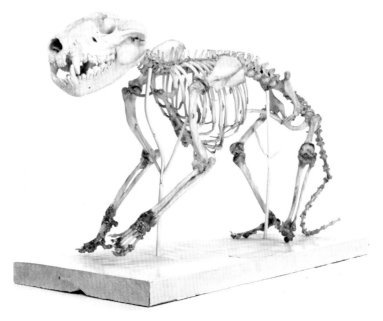

Wildlife diseases tend not to get as much attention as other factors in species decline and extinction, but they are increasingly being implicated in the demise of a large number of species. Conservationists have long been concerned about populations that become isolated in small pockets of habitat or have low levels of genetic diversity as this can make them susceptible to emerging diseases. In these situations it doesn't take much for a disease or parasite to take hold of the whole population, as all the individuals have very similar genes. If a disease finds a chink in the armour of the immune system in one individual, it is likely to be able to infect all of them.

This is what happened to the Tasmanian devil. Although their population was large before 1996 – estimated to be 130,000–150,000 – they didn't show a lot of genetic diversity. This is possibly because only a small number of devils originally colonised Tasmania, and all living devils are their descendants, with their same genes. In any case the fact that they were all genetically very similar left them susceptible to a new disease, and the one that appeared in 1996 could hardly have been more devastating.

This was the year that devil facial tumour disease (DFTD) emerged. It was first discovered near the north-eastern tip of the roughly heart-shaped island, and in the subsequent 20 years it has spread almost across the whole state (complete coverage is anticipated very soon). When a devil contracts DFTD it is virtually 100 per cent fatal, and unfortunately most devils catch it. In some places populations have crashed by more than 95 per cent, but overall numbers are down 80 per cent across its entire range – an extraordinary crash in just two decades.

DFTD is a cancer that is transmitted from one devil to another. It affects the face and mouth: the tumours essentially eat away at their facial tissues. I have been on a number of fieldtrips to catch, study and release devils with the scientist Dr Rodrigo Hamede working on the disease at the University of Tasmania. Despite having encountered a couple of hundred devils, it is still very upsetting to find devils with much of their face and teeth missing – they are clearly in a lot of discomfort.

Contagious cancer is a terrifying thought, and there are only a couple of known instances in nature. DFTD is spread by the transmission of the cancer cells themselves. Rather than a bacterium or virus, the cancer cell itself is the infective agent. When infected devils bite another devil and a cancer cell rubs off, that devil is infected. In other words, clones of the same cancer cell that arose in the first devil around 1996 (patient zero) is still infecting devils nowadays. As Hamede says, the disease has managed to become immortal, just as long as there are devils ...

As devils bite each other a lot, in feeding, mating and saying hello, death is more or less guaranteed. Given the importance of this species to the island's ecology (not to mention the tourism industry), scientists

have been prioritising DFTD research since it appeared. Success has not been forthcoming and the disease is spreading and devils are dying.

In 2016 two studies were published resulting from the University of Tasmania's fieldwork which suggest that the devils themselves are just starting to fight back. First, they found that the corresponding genes which in humans relate to immune responses to cancer are changing in devil populations. This doesn't mean that they are evolving immunity (yet), however given that devils will be under serious pressure from natural selection to fight the disease, it does show that evolution is taking place. This is amazing as it has happened in just four to six generations since the disease appeared, which is incredibly fast.

Second, some devils are now being caught with tumours that shrank over time – this is unprecedented as typically tumours grow until the devil dies. I was actually with the team that caught some of the first of these devils in the field. It was pretty exciting when we checked the tumour measurements against the ones taken the previous season, and the blood taken from these animals was found to have antibodies against DFTD. It's far too early to get carried away with suggesting devils are on the road to recovery, but again it shows that evolution is in action.

In the meantime conservationists are attempting to secure 'insurance populations' outside of their native range in case they do become extinct on Tasmania, so they could be reintroduced. This isn't straightforward, as introducing a new predator to an area is almost certain to have an impact on the existing wildlife. I visited Maria Island, naturally devil-free, but within sight of the Tasmanian mainland in 2010 and again in 2016. In between that time a population of disease-free devils was introduced.

The devils were doing fantastically well – I encountered them a number of times in the wild there (something that only rarely happens on the mainland, though this does suggest they were less wary) and their poo was everywhere, indicating a very healthy population. However, on close inspection their poo was full of penguin feathers, and now the once common penguins of Maria Island have virtually disappeared. They are found in many other places, so this will not have an effect on their global population, but it does serve to remind us that sometimes we have to prioritise the conservation of one species at the expense of another.

80 Tree Frogs

PRESERVED SPECIMENS

Extinction And Wildlife Disease: Fungal Frogs

Tasmanian devil facial tumour disease is thankfully specific to that one single species of carnivorous marsupial. However, there are diseases which are far more wide-ranging. In 1999 a disease was described that threatens an entire vertebrate class. In recent years amphibian populations around the world are being wiped out by a parasitic fungus called *Batrachochytrium dendrobatidis*, belonging to the chytrid family. To date

hundreds of amphibian species have been driven to decline or extinction by this disease. Its impact has been unprecedented.

In case you think I'm overstating things, a 2007 report by the IUCN (World Conservation Union) described chytrid as 'the worst infectious disease ever recorded among vertebrates in terms of the number of species impacted, and its propensity to drive them to extinction'.* No other known disease has ever had such an impact on global biodiversity.

The impact can be immediate – with populations declining just a few weeks after the disease arrives. The disease caused by chytrid affects

---

* Gascon, C., Collins, J.P., Moore, R.D., Church, D.R., McKay, J.E. and Mendelson, J.R. III (eds), *Amphibian Conservation Action Plan* (IUCN/SSC Amphibian Specialist Group: Gland and Cambridge, 2007)

amphibians' skin, which is a highly sensitive organ. Most terrestrial vertebrates gain their nutrition by eating and drinking, and most of their oxygen by breathing (we fall in to this category, obviously). Amphibians, by contrast, do a high proportion of these functions directly through their skin. This means their skin has to be tough enough to survive the rough and tumble of everyday existence, but thin enough to breathe and 'drink'.

This balance is achieved by a layer of cells that produce the fibrous protein keratin, making the skin resistant to damage. This is the same tough material that forms hair, nails and reptile scales. Frogs' bellies have an area of skin that is responsible for a lot of the absorption of water and salts – it's called the drink patch.

If an animal is infected with enough of the chytrid fungus, it causes the skin to produce too much keratin, which stops the absorption of salts through the skin, particularly through the drink patch. Without electrolytes like calcium and potassium muscles can't function, including the heart: infected frogs die because their hearts stop beating.

Although it was only described in 1999, it is clear that this species of chytrid existed in some amphibian populations in different parts of the globe before that time, but now it is found in every continent where amphibians live. While we don't know when or where chytrid first caused population declines, it is clear that humans are responsible for spreading it around the globe. Amphibians are transported for human purposes for food, as pets and for use in lab experiments, and this is how chytrid has been spread to every corner of the amphibian earth.

Once in a new environment the fungus can spread through spores very easily – either directly through contact from one frog to another, or simply by drops of infected water making their way into uninfected water courses or even just a damp environment. If an animal visits one stream or pond, it can carry spores to the next place it visits just by getting wet.

Today is not a good time to be an amphibian. The IUCN, who are responsible for categorising the global conservation status of species and ecosystems, estimate that 42 per cent of the world's amphibian species are threatened with extinction, and that number may be as high as 56 per cent. More than half of the known species of frog, salamander and caecilian could be dying out. That is an incredible statistic. While chytrid obviously causes a devastating disease, it isn't even the biggest threat to amphibian survival. That 'award' goes to habitat loss – humans keep destroying the places that they live.

81 Marine Iguana

SKULL

Extinction And Climate Change

It is now effectively beyond doubt that human activity is driving an unprecedented change in global climates. Carbon dioxide and other greenhouse gas emissions are raising the earth's temperature at an alarming rate. For human populations, this is having a frankly terrifying effect on food production and the availability of freshwater, as well as the risk of our settlements being inundated by rising sea levels. The impact on global biodiversity is no less severe, and it's not difficult to understand why.

The distribution of natural ecosystems around the globe is dependent on the complex interactions of a wide number of different environmental variables. For example the specific communities of plants which form the basis for a given habitat can only survive in a specific range of

temperatures, rainfall levels and soil types. Climate change can alter all of those factors (and more), and therefore impact all of the species that rely upon the habitat for survival.

Climate change is shifting the occurrences of ecosystems around the globe (but not evenly – there will be a significant relative increase in the amount of desert). In order to stay within the range of the environmental conditions they are adapted to, many species are being pushed to higher altitudes (up mountains) or to higher latitudes (towards the poles). It might be tempting to view this as a potential positive outcome for local biodiversity. There are species of European butterfly that are extending their range into the UK, for example, which had previously found it too cold and wet.

However, in the main this is a very worrying trend. Many species are on landmasses at the northernmost or southernmost limit of their range, and they can't go any further towards the pole as there is no more land. Likewise, there are species that cannot go any further uphill, as there is no more mountain to climb. The ptarmigan is a species of grouse whose British range is restricted to the mountains of Scotland. These mountains are not very high, so as temperatures warm, their snowy habitat is completely disappearing – they cannot go any further.

Polar bears are the poster children of climate change. They spend the winter on land, barely eating for five months. As soon as the sea begins to freeze in the autumn, they can return to the ice to hunt seals – an activity they are perfectly adapted for. However, climate change is causing the sea ice to form later and later in the year. The fat supplies that the bears put down the previous winter will not last over the lengthening ice-free seasons, and they will starve before the ice forms.

While many species' populations are being affected by climate change, it appears certain that human-induced climate change has already driven many species to extinction, many of which have never even been discovered. However the first reported species of *mammal* to be completely driven over the edge by climate change was the Bramble Cays melomys, in 2016. This small tropical rodent was native to an island at the northern end of the Great Barrier Reef, off the coast of Queensland, Australia. Its entire population was wiped out by rising sea levels, caused by the melting of the polar ice caps.

Despite the fact that extremely few people had ever heard of a melomys (even though several species live in Australia), the story of this landmark extinction made international headlines. And the polar bears' situation is very well known. Mammals get a lot of coverage, but climate change is going to have a far wider impact than on this small group of charismatic vertebrates.

A study in 2015 demonstrated that cold-blooded species (more accurately called ectotherms: those which do not generate their own body heat and have to absorb it from the environment) are not very tolerant of rising temperatures. Each species of ectotherm will have a specific range of temperatures within which it can survive. Outside of that range their bodies shut down and they die.

Ectotherms can gain or lose heat through their behaviours, such as basking in the sun, panting, swimming in cool water and seeking shelter, but if their environment changes to such a degree that they cannot escape the heat, their internal systems cannot cope. The researchers found that this was particularly true of species that live on land. Therefore global warming will affect *huge* numbers of species in this way – reptiles, amphibians, and the largest animal group of all: insects.

As temperatures increase, species can either move towards the poles or to higher altitudes, spend more time in the shade or evolve the ability to cope with higher temperatures. However climate change today may be happening too quickly for species to adapt.

One ectothermic species that has been formally listed as Vulnerable to extinction is the marine iguana – an enigmatic lizard from the volcanic Galápagos Islands that lives on the coast but regularly swims out to sea to graze on underwater algae. This species is also known to be significantly affected by El Niño events. This is a periodic natural weather pattern involving the failure of the trade winds that 'normally' blow westwards across the Pacific, occurring every 2–7 years. These winds push the warm surface waters of the eastern Pacific (off the South American coast) across the ocean, resulting in cool water welling up from the depths and replacing it. During an El Niño year, when the winds fail, the temperatures around the eastern Pacific where the Galápagos Islands are found, can soar dramatically.

This has historically been a natural occurrence, and once iguana numbers have crashed following a typical El Niño year, populations can generally recover as competition is lower. However, climate change is now understood to be causing an increase in the regularity of 'Super El Niño' events, when the impact is far greater, possibly even doubling their rate.

Combined with the other threats to marine iguanas – introduced cats, dogs and rats; disease (particularly stemming from the influx of people through tourism) and oil spills – this all adds up to a worrying time for the islands' iconic species.

A Galápagos marine iguana.

82 Crescent Nailtail Wallaby

PRESERVED JOEY (POUCH YOUNG)

Extinction And Introduced Predators

You may have noticed that there is a rather Australian bias in this Natural Histories section of this book. On the whole, I don't really have an excuse for this – it's because I love Australian wildlife and I want other people to love it too. Australia has some incredible, unusual and unsung species, which as I've mentioned sometimes get put down by some people who think them inferior to animals elsewhere in the world, which is absolute rubbish.

The Australian bias is particularly apparent in this segment about extinction. However, that's not just because my particular zoological interest is Australian mammals, but because arguably there is nowhere on earth that has suffered such a catastrophic species loss in such a short space of time. As a direct

Crescent nailtail wallabies are one of at least twenty-nine species of mammal that have disappeared from Australia since European invasion.

result of Europeans invading Australia in 1788, at least 29 of the 315 native land mammals there have been driven to extinction, and by far the majority of the remaining species have suffered a dramatic population crash and range restriction. That is an extraordinary impact in a very short space of time.

This object, the crescent nailtail wallaby, is here to demonstrate the statistic that I find singly more remarkable than any other number in biology ... Ready?

It is estimated that 31 million native animals are killed in Australia by feral cats *every single night*.

This number comes from combining a rigorously calculated estimate for the number of cats in Australia, which was assessed to fluctuate between 2.1 million after a severe drought and 6.3 million after extensive wet periods. This was the result of a 2016 study involving a significant proportion of Australia's leading mammal ecologists, who also concluded that cats are present in over 99.8 per cent of Australia's landmass. The population count can be combined with the number of small mammals, frogs, lizards, birds and snakes cats kill each night. The stomachs of captured Australian feral cats are routinely found to contain between five and thirty freshly killed natives.

Crescent nailtail wallabies were a small hopping marsupial, named for the moon-shaped shoulder marking and the horny 'nail' at the end of their tails. They lived across a massive area of arid and semi-arid Australia, ranging over much of the south, central and southwestern parts of the country. They were common when they were described in 1841. They were extinct a little more than 100 years later: the last accounts are from before 1950.

Cats and foxes, which are not native to Australia, were to blame. On the whole, Europeans have made Australia a pretty terrible place to live if you are an Australian mammal, which is particularly unfortunate as that's the only place that nearly all of these species do live. The cats and foxes they/we introduced are arguably the country's worst destructive agents, but there is a long list of other factors, including: dramatically altered fire regimes; introduced pigs, mice and rats (which have been most impactful on ground-nesting birds); habitat destruction for towns and agriculture; introduced herbivores (including rabbits, sheep, goats, horses, donkeys, cows, two species of wild cattle, camels, hare, six species of deer); diverted and drained rivers and wetlands; climate change; dredged sea-grass beds; polluted coral reefs; and introduced cane toads (see chapter 67).

Elsewhere in the world, the risk of going extinct increases with body size: bigger animals are more likely to become extinct due to human activities. This has a lot to do with the fact that large animals generally take longer to reach reproductive age, and breed less frequently, which means

A feral cat in Australia, with a dead brush-tailed phascogale – a native marsupial carnivore.

that their populations can't readily bounce back. Other factors play a part too, like humans being more likely to kill big animals for food or sport.

Australian extinctions, however, follow a different pattern. There, modern extinctions have particularly affected species in what's called the 'critical weight range', from about 35g to 5.5kg. The reason is that this is the ideal size of animals to be hunted by foxes and cats (at least when they're young, for species at the upper end of the scale). Species that are smaller than 35g reproduce quickly enough to absorb some of the damage done; larger species are too big for cats and foxes to tackle. This effect is most strongly seen in drier parts of Australia, where the vegetation gives small animals few places to hide; and with species that live on the ground, where cats and foxes hunt.

It's no surprise that cats were largely brought to Australia as pets, as well as for controlling the rats and mice we introduced. However what particularly irks me is the reason that Europeans introduced foxes. Even though the country was full of native species that could be equally easily chased around with a pack of dogs from the back of a horse (not that I would condone that for a second), foxes were let loose in Australia so that the settlers could enjoy a bit of fox hunting. It beggars belief.

Extinctions and population crashes caused by non-native species are by no means a problem restricted to Australia. Native species evolve adaptations best suited to the places they live, and that includes forming defences against their *natural* predators. When new predators arrive in large numbers, delivered to new lands by people, the results are usually devastating. Cats, rats, pigs, weasels and dogs are the worst culprits, and islands usually get the brunt of it (these are places where the animals tend to be more naïve, having evolved in the absence of predators). Dodos are of course the classic example, which suffered most from rats and pigs eating their eggs. It is a global problem. In 2016 a study found that introduced predators have contributed to more vertebrate extinctions since 1500 than any other cause. Of the extinctions of birds, mammals and reptiles in that period, 58 per cent can be blamed, at least partly, on introduced mammalian predators.

83 Trilobites

TWO NEAR-COMPLETE FOSSILS

'The Great Dying' And Mass Extinctions

I'm sorry that the last few chapters have been a bit depressing – it's hard to talk about man-made extinctions in an upbeat way. It's true that the heroic and thoroughly underfunded efforts of the modern conservation movement have had some small and possibly even medium-scale successes, but it's hard to argue that the outlook isn't bleak for the current extinction crisis.

Dalmanites sp. (*left*) from the Silurian, around 430 million years ago, and *Scutellum sp.* from the Devonian, 420–360 million years ago.

To look on the bright side, it could be even more unpleasant: there have been at least five points in global history when it was a *worse* time to be alive. In that sense, animals today have it lucky(!). The scale of biodiversity loss today is being described as the sixth mass extinction. Species go extinct all the time, but a mass extinction is when the rate of extinction significantly outweighs the rate of speciation over a brief period of time. This results in sharp declines in the global biodiversity.

The definition of a 'mass extinction' isn't particularly specific in terms of how fast or how massive an extinction trend has to be, but it is generally held that the fossil record of complex life (it's pretty much impossible to assess changes in the diversity of fossil bacteria) has experienced five 'major' mass extinctions to date.

The most famous mass extinction happened around 66 million years ago, as the Cretaceous transitioned into the Palaeogene. This is the point at which the non-avian dinosaurs, pterosaurs, marine reptiles, ammonites and other museum favourites disappeared. However, it was by no means the most devastating. That accolade belongs to the End Permian Extinction, also known as 'The Great Dying'. There has never been a worse time to be an animal than at the end of the Permian period, 252 million years ago.

Picture a packet of biscuits with twenty-five biscuits in it. That packet represents global marine diversity during the Permian period – the last stage of the Palaeozoic era. Move forward to the beginning of the Triassic (the first period in the Mesozoic era) and there was only one biscuit left: during The Great Dying, 96 per cent of marine species disappeared. It was literally the end of an era. Truly, if it had been any worse there would have been nothing left to rebuild from. Game over.

Things weren't much better on land, where 70 per cent of vertebrates died out, along with many other things – nearly all the trees disappeared, for example. It was also the only mass extinction to affect the insects – usually the masters of survival. Eight insect orders became extinct during the End Permian Extinction, compared to only two orders that have disappeared in the quarter of a billion years since.

Among the casualties were many of the therapsids, but evidently they weren't all wiped out as they gave rise to us. Amphibians and their relatives also suffered heavy losses. Pretty much every lineage was massively reduced in diversity and some significant groups disappeared altogether – the most famous of which were the trilobites.

Trilobites had been a major factor in marine ecosystems for around 270 million years before they were wiped out in the Great Dying (though they had been decreasing in diversity for some time before the big event). They were among the earliest fossils known with complex visual systems, with eyes containing huge numbers – sometimes 15,000 – of individual

A close-up of one of the huge wrap-around eyes of *Dalmanites*.

lenses. These lenses, which in at least some species were made of pure crystals of calcite (a colourless form of the mineral that forms limestone) focussed light on receptor cells below. This gave them incredible abilities to detect movement – critical for both avoiding predators and finding prey.

During their 270 million year history, trilobites evolved a huge diversity of variation on the 'looks-a-bit-like-a-fancy-woodlouse' theme. Some could roll into balls, some had spines, and some had elaborate projections forming shovels, prongs and completely unknowable processes sticking out of their exoskeletons. They filled many ecological niches, including filter-feeding, scavenging and active predation. Some moved on the sea floor, some just below it, and some swam in open water.

The trilobite depicted here – *Dalmanites* – is from the Silurian period, around 430 million years ago. This species' most distinctive feature is its eyes (only one survived on this specimen, but it's been nearly half a billion years so let's not be too hard on it), appearing like curved digital advertising boards wrapped around a tower. Most of the individual elements of the compound eye have been preserved here, and while it may not have had the vast number of lenses of some of its relatives, the convex curvature and the positioning would have given it 360-degree vision. Their towering nature suggests that this species lived buried under the sediment, with its eyes poking out into the water above.

Trilobites did not survive the End Permian Extinction, along with *most other things*. The cause of the world's greatest catastrophe is not precisely known. It appears not to have been particularly sudden, possibly occurring in a couple of bursts of extinctions over a few million years, and a number of factors are likely to have contributed. One of the likeliest candidates is volcanic eruptions in Russia – there are 2 million sq. km of lava of this age in the Siberian Traps.

The event was associated with a giant leap in global temperatures, tied to a huge increase in carbon dioxide levels (something that will give us little comfort today). Dissolving in sea water to form carbonic acid, this would have stopped hard-shelled marine species from forming their calcium carbonate coverings, which dissolve in acid. This helps explain why life in the sea all but ended 252 million years ago.

Those few animal groups that survived took a long time to fill the niches left vacant by The Great Dying, and the fossil record of the early Triassic is sparse as a result.

84 Woolly Mammoth

HAIR FROM FROZEN SPECIMEN

De-extinction

Extinction is forever. I can't really think of a stronger three-word sentence. But is it true? In the past couple of decades, the idea of bringing extinct species back has gained momentum, and the technology behind it has developed at an astonishing rate. We're not there yet, but in the near future it may be possible to genetically recreate a healthy individual from a once lost species. It has become known as de-extinction.

Significant genetic advances have been made that could take the genetic make-up of an extinct species and use it to 'de-extinct' them. If cells from extinct species could be found which still had viable genetic information in them, could they be cloned by injecting the genetic information into a surrogate egg from a related species? This has already been done. Nearly.

In 2000 the last Pyrenean ibex – a kind of wild mountain goat – was found dead, crushed under a fallen tree in her Spanish mountain home. However, some months earlier she had been briefly captured and some skin cells were collected and frozen in liquid nitrogen, preserving her genetic information.

After her death, these cells were used to clone her using goat surrogates (surrogoats?) to carry the developing embryos. One baby Pyrenean ibex clone was born – it was the first de-extincted animal in history. However, it died after just seven minutes as its lungs had failed to develop properly.

Genetic technology has come on significantly since then, using developments from genomics, IVF, gene editing and stem cell research. There are a number of ways that these innovations could be applied to extinct species. If the genome – the exact arrangement of the DNA code of a species – could be established on paper, could chromosomes be built synthetically to recreate them in a test-tube, and then inject them into a surrogate cell and bring them back?

In reality this is not currently possible – the idea of taking a string of letters (representing the DNA code) on a computer screen and turning it into a functioning animal chromosome has never been achieved, let alone for an extinct species. The next problem is that the genome of an extinct species has never been fully sequenced – DNA degrades very quickly over time and is not preserved well in museum specimens.

The passenger pigeon is the poster child for de-extinction. This North American species was once the most abundant bird on earth, forming flocks of 1 billion individuals. Despite its vast numbers, a century of hunting for meat (it was sold by the ton) resulted in an unprecedented population crash. The last individual died in 1914. Could the passenger pigeon be brought back?

People are working on it. One option is to compare the genomes of passenger pigeon specimens in museums with their closest living relatives, the band-tailed pigeon. Since closely related species already share huge portions of their genomes, could gene editing technology allow the extinct species' genes to be substituted into the surviving relatives' chromosomes at the points at which they differ?

Hair from a mammoth carcass found in the permafrost of far eastern Russia, pictured with a model mammoth.

This question is also being asked of an animal that has been gone for a lot longer than one century: the woolly mammoth. Incredibly, entire mammoth carcasses complete with skin, muscles and organs are found emerging from the ground as climate change melts the permafrost – the Arctic soil that has remained frozen since the last ice age. This hair came from such a carcass. Could we learn enough about the mammoth genome from the fragments of degraded DNA in these specimens to edit the genome of an elephant to recreate a mammoth? As it stands we are still a very long way off knowing what made a mammoth, genetically.

There is also the question as to whether all of this is a good idea. There's a famous quote that springs to mind: 'scientists were so preoccupied with whether or not they could that they didn't stop to think if they should.'*

There are a number of motivations behind the push for de-extinction. It is our fault that species like the thylacine, crescent nailtail wallaby, passenger pigeon and (probably in part) mammoth are gone – doesn't that mean that it's our moral responsibility to bring them back? Then there's the sense of wonder that we could experience by encountering such species in the wild. And from a conservation point of view, some of these animals were keystone species in their habitat so perhaps bringing them back would bring back the entire ecosystems that they helped maintain.

Aside from extinct species, these technologies are also being explored to boost the genetic diversity of critically endangered species that are still with us, for example by inserting genes for resistance against threatening diseases. These are some very interesting developments.

But there are serious arguments against de-extinction. How could they ever recreate a self-sustaining breeding population? Not to mention the ethical considerations of forcing a surrogate mother of a different species to carry the embryo. For the mammoth, that would be an Asian elephant, which means a cruel twenty-two-month pregnancy.

I can't think of a worse option for de-extinction than the passenger pigeon – perhaps the most social creature ever. It went extinct because it required gigantic flocks to stimulate ovulation – how could genetic engineering ever recreate that? Even if we did, does the habitat still exist to support our lost species?

The biggest problem with de-extinction is a political one. It sends the message that extinction *isn't* forever, and that could do a lot of damage to the conservation movement. As it stands, the technologies are close to existing to bring something back *in theory*, but we don't have the information about the extinct genomes to actually do it yet in practice. It is an interesting time to be extinct.

---

* *Jurassic Park* (1993), directed by Steven Spielberg [Film]. USA: Universal Pictures

85 Collie and King Charles Spaniel

SKULLS

Domestication

De-extinction would be by no means the first time that humans had genetically manipulated an animal. The process of domestication involves the separation of some members of a species from their wild-living ancestors. In many cases this has resulted in profound differences between the two groups, both genetically and anatomically.

While humans aren't the only species to engage in domestication (remember the fungus-farming ants, chapter 72), we are surely the one which has exerted control of the widest number of species. Definitions of domestication involve one organism having long-term influence over the reproduction and care of another organism for a number of generations, for their own benefit.

That benefit most obviously manifests in the animals we farm for food. We've also harnessed the natural power of species to help us create or catch food – the domestic honey bee is one of our most important 'tools'

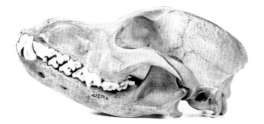

in some parts of the world because of its pollination skills; and dogs, ferrets and falcons have been used to assist in hunting. Animals are bred for their skins and many other products; large animals have been bred to be beasts of burden and as modes of transport; and of course we've taken animals into our homes for our own pleasure, as pets.

Humans' first domestication was that of the dog from the grey wolf, and current evidence suggests that this happened earliest with European hunter-gatherers between 19,000 and 32,000 years ago, though it's perfectly plausible that wolves were domesticated more than once over their vast range across the Northern Hemisphere. It likely happened like this ...

Some wolves followed nomadic encampments and scavenged off of their waste (as they do around farms, towns and cities today). These animals also would have benefitted from our hunting behaviour, by bringing down prey animals that had been injured or separated but yet escaped a human hunter. Such behaviour is a common occurrence where two predatory species share a range. One group goes in for the kill, while the other waits on the side lines for an opportunity to jump in and take advantage of the carnage.

Why did the humans put up with the wolf at their door? They must have benefitted from the relationship. A likely early positive outcome was the vocal warnings against approaching predators or human rivals that the nocturnal wolves provided. This didn't make them man's best friend straight away, but they were at least a helpful neighbour.

This initial behaviour – where some individual wolves were associating themselves with hunting camps – started the process of reproductive separation from independent wolves. Two diverging gene pools (those packs that followed humans and those that didn't) with little mixing, allowed for the process of genetic isolation which began the domestication of wolves.

The humans, it is believed, then directly interacted with these friendlier wolves, by taking pups into their societies and raising them. Those individuals that did not run away – and furthermore allowed this interaction without attacking – would have been bred together by the people. This was the first time that animal reproduction was controlled by humans.

In support of this, during the early contact period following European arrival in the New World, Native Americans and Amazonians were observed regularly bringing 'wild' animals into their homes and villages and caring from them. This was especially common with regards to younger animals, in particular bears, and women were even reported to breastfeed these pets.

Wolves are particularly suited to their intimate level of domestication because of the similarities in the complex social structures of both humans and wolves. Although many species of canid (the dog family) are social pack hunters – particularly coyotes and African hunting dogs –

none has the highly structured hierarchical dominant/submissive pattern of the grey wolf

Once in the community artificial selection could take place, where humans determine which characteristics are going to proliferate by favouring individuals with desirable behaviours. Anatomical changes as well as behavioural ones took place, as genes are not independent of each other. When you select genes that control a certain characteristic – like aggression – you automatically select for those genes that are situated near them on a chromosome, or other characteristics that are related to the features you are promoting.

Wolf pups would have benefitted from human communities if they had the ideal combination of favoured physical attributes and the temperament to survive to breeding age among people. Continued filtering of cubs with this mixture of traits would have inevitably led to animals which differed from the wild form in both morphology and character. Those animals were dogs.

As to why the morphological differences we see between wolves and domestic dogs occurred – features like shorter snouts with smaller and more cramped teeth and floppy ears – it is credibly suggested that these characteristics are favourable by people as they are 'cute'. In fact they are features seen in infant wolves, and thus perhaps 'babyishness' is more tameable. An interesting experiment in Russia led by Dmitry Belyaev tested how such changes might come about, by selectively breeding foxes with the single target of making them more tame. After more than 40 years of breeding together the tamer individuals, the foxes they produced did indeed carry the same traits by which we identify domestic dogs – short, curled tails, short legs, floppy ears, and a lighter coat, even though he was not deliberately trying to select these features.

Once man and wolf-dog could exist together happily, other benefits were realised. Wolves are very nifty hunters, and humans harnessed this skill in their new companions. It wasn't until the last few thousand years that individual breeds with different suites of characteristics were developed. Today, humans have forced the descendants of wolves to become the most anatomically variable of all species. The two drastically different skulls pictured on p. 267 are both dogs.

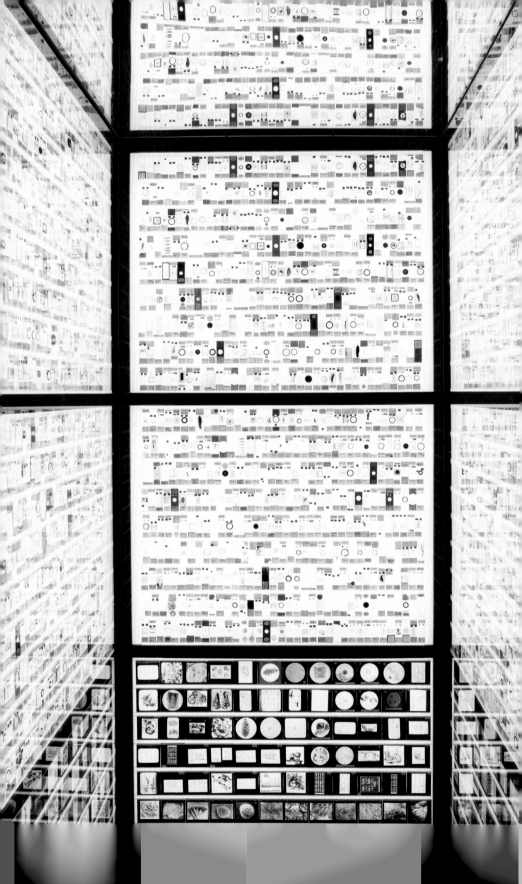

Part 4

Displaying Nature

I manage the Grant Museum of Zoology at University College London. As a collection founded to teach evolutionary principles in 1827 (32 years before Darwin published on the topic), the specimens in the Grant Museum have always been arranged taxonomically. It's an unnatural way of presenting them because although they belong to the same mammalian order, lions would never be seen with walruses, for example, outside of a museum.

In placing them together we focus on one aspect of the way we see them: through an evolutionary lens. But in so doing we strip these species of a lot of their essence of being. Yes, lions and walruses have some shared anatomy resulting from their shared ancestry, but what does that really tell us about them? In the museum they stop being wild animals and become static artefacts arranged in our chosen human system. It is an exercise in both comprehension and control.

And it's easy to picture animal specimens in a museum as truthful representatives of their species – however, skeletons are wired together and taxidermy is stretched into position by people. Although these specimens are *made of* the animal itself, and so in a sense are 'real', they are also manufactured objects. Such constructions come full of the biases and misunderstandings, and sometimes political motivations, of the people that preserved and commissioned them.

In this final section of the book, my last fifteen objects explore the ways that museums represent nature. We have already seen how one central branch of natural history – taxonomy and classification – is an artificial structure through which we comprehend the enormity of the animal kingdom. In the following chapters I'll be considering the ways in which we display and communicate nature, which is similarly synthetic.

Natural history museums are places for people, made by people. We might like to consider them as logical places, centred on facts, but they can't tell *all* the facts – there isn't room. We museum professionals make decisions about what to say and what to put on display based on what visitors want to see, and also what we want to show them.

This is very much a personal take on the nature of natural history museums, and in no sense is it a criticism of the sector to which I have dedicated my career and most of my time. What I hope to unravel is the means to think about natural history museums as a product of their own history, and that of the modern societies they are embedded in. They are not apolitical, and they are not entirely scientific. Like taxonomy, they are a non-natural framework we use to understand the world we live in.

The following chapters will explore the theme of how we see the natural world by investigating how museums came together, and how specimens come to be preserved and displayed. The techniques used rarely avoid manipulation and interpretation of the dead animal, and what results may not be a true depiction of the species they were intended to represent. And that is what natural history museum specimens are for: specimens are *examples* of a wider thing. They are there to *represent* something bigger. A taxidermy platypus in a museum is not actually a platypus, but it is made of platypus. In a display case, such a specimen is intended to represent all of platypus-kind, and depending on the display, perhaps all egg-laying mammals, and maybe all Australian animals. Unlike objects in museums of most other disciplines, the very nature of a *specimen* is to be a typical *example*. Although in reality they may not be.

My hope is that when people visit museums they may be able to consider the human stories behind the displays they see. They might consider the question of why is all that stuff there: what is that museum – or that specimen – doing? What is it for?

We will explore some of the practices of natural history in museums, such as how species are described and how we decide what to display and what should stay in the storeroom. A couple of animals get special attention here – dinosaurs and dodos – two mainstays of natural history museums. They have stories of their own, and the way they are represented in museums provides interesting windows into the ways we think about nature more broadly. I will also consider the ways that humans are incorporated into museums, and whether the boundaries between natural history and art really exist.

To end, I will argue that despite these potential flaws or foibles, natural history museums are essential for us to appreciate, communicate and understand the natural world. Their contributions to world-changing science cannot be overstated.

Natural history museums are an essential route for many people to be inspired by the natural world. Most zoologists – myself included – would attribute at least some of their 'calling' to time spent in museums. And museums are wonderful – often deliberately founded and constantly striving to engage people in the natural world. But sometimes it's easy to forget that natural history is inherently unnatural.

86 Domestic Pig

PRESERVED CONJOINED SPECIMEN

Cabinets Of Curiosities Are Dead. Long Live Museums.

If we were to seek the origins of the idea of the museum, we arrive at the cabinets of curiosities of Europe in the fifteenth to seventeenth centu-

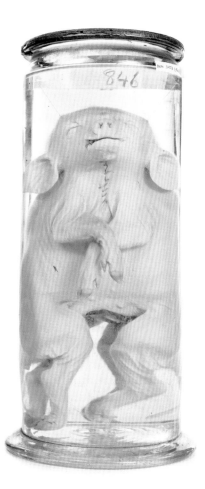

ries. These themselves were the conceptual descendants of religious treasuries that kept (or claimed to keep) sacred religious artefacts and relics such as splinters of the True Cross or some finger bones of a saint.

Personally, I have some issues with the way in which 'cabinets of curiosities' is sometimes banded about in relation to modern museums that I'd like to explore here – I don't think the connotations of these words do our museums any favours. It is certainly true that the collections of a number of the world's oldest museums were originally amassed as true cabinets of curiosities, but I don't think this is the way in which we should consider them today.

Pictured on p. 275 is the earliest known image of a cabinet of curiosities, that of Ferrante Imperato in Naples in 1599. Imperato was an apothecary (which is a common trend among cabinet-keepers from this period) and he would have been able to use his collection for research and in the attempt to develop new medicines.

What we can see from the image is that research was not his only motivation. Unlike the sacred treasuries that came before them – which kept their objects locked away, safe from harm and from visitors – cabinets of curiosities were intended to be viewed and enjoyed (albeit by the very highest levels of society).

Filled with works of art, shells, seeds, minerals, animal skins and skeletons, preserved specimens, medical and zoological 'abnormalities' (like the pig foetus on the previous page with one head and two bodies), clockwork automatons, scientific instruments and ethnographic objects from around the world, cabinets of curiosities were a symbol of power, status and sovereignty. The collections were arranged and categorised in ways we would hardly recognise today. Different collectors would have had different interests, however. For example we can see that Imperato was clearly keen on the '*naturalia*'. Cabinets of curiosities sought to create a microcosm, fitting the whole universe into a single room, distilling all human knowledge into a collection of wonders.

There are significant differences between the cabinets of curiosities of the Renaissance and the modern natural history museum, and not just because we no longer hang our crocodiles from the ceiling (this became a common feature of the cabinets, perhaps stemming from this engraving of Imperato's collection, but there are also examples of crocodiles hanging from the ceilings of European churches from the same period).

When museums today have displays that are very rich in specimens, it is common to hear things like 'Isn't that museum wonderful! It's such a cabinet of curiosities!' This exclamation is clearly meant as a rich endorsement, obviously intended as a compliment. Nevertheless, it makes me wince.

Inspiring curiosity and wonder is surely among the highest ambitions a museum could ever have. It's infinitely more important than making visitors learn something. The best museums encourage curiosity, and they obviously have cabinets, but that does not make them cabinets of curiosities.

For me, the risk is that using this term today implies that our objects are nothing but curios – weird artefacts amassed by some eccentric collector. Erratically accumulated in another time; weird and wonderful titbits intended to impress; to show off the collector's status and influence – 'Gosh, Sir William! Where did you get that ghastly tenrec!?'

Not to say that other disciplines do fit that description, but I think the phrase cabinets of curiosities – and *Wunderkammer* like it – are especially bad fits for natural history. It's not that I'm saying we should be particularly pious about the real-world contributions that natural history collections have made to support world-changing science (of which there are many). I'd be the first to admit that dead animal specimens can be funny, and beautiful, but they are also important.

These terms suggest that our collections are irrelevant today – nothing but a sophisticated side-show. Perhaps not even that sophisticated.

Ferrante Imperato's cabinet of curiosities, 1599.

I'm aware that not all of my colleagues agree, and some have even embraced it. I understand that for some it might feel like an attractive proposition, but I do think it undermines the importance of museums.

To be clear, I absolutely subscribe to our collections being objects of wonder, but I'm left very uncomfortable at thinking we are working so hard to attract people to ogle at curios. This is one of the reasons that I don't put specimens like the conjoined pig on display in the Grant Museum – I'm not confident it won't be looked upon as a 'freak of nature', whatever we may say about it on the label.

Another word that museum professionals avoid is 'weird' (and I know at least one museum's communications strategy explicitly forbids it), for entirely the same reasons – museums shouldn't be getting public support to maintain curios and weird trinkets.

In many ways Renaissance cabinets of curiosities were the precursors to museums. They held art, zoological and geological specimens, archaeological and anthropological finds and other travellers' spoils from across the globe, largely before any of these academic areas really existed.

But today these disciplines do exist, and we do use our collections in interesting, valid, useful, relevant, important, wonderful ways. So I ask you a favour, please don't think of our museums as cabinets of curiosities – I think we are so much more than that.

87 Entomology Collection

DRAWER OF PINNED ARTHROPODS

How Did Museums Get All This Stuff?

Most natural history museum professionals will tell you that there are two questions that we most regularly get asked when talking to visitors about our specimens. The first is 'Is it real?' The answer to that is nearly always yes (the exceptions will be discussed in the following chapters), the chief reason being that museums are exactly the places you should expect to

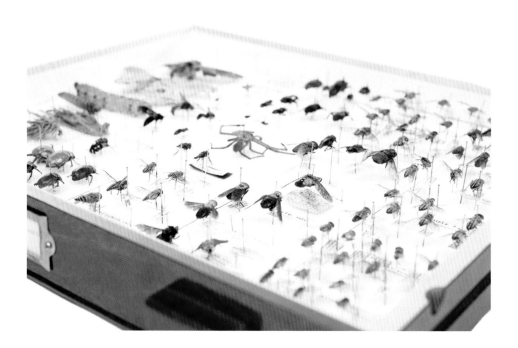

find 'real' objects, and that in most cases it was a lot easier to collect the genuine article than fabricate a realistic replica.

The second is 'Did you kill it?'

For clarity, most of these visitors are genuinely asking if the person they are talking to killed the animal *themselves*, rather than enquiring as to whether the specimen was killed for the museum at some point in the past.

Although some curators at museums which are also research institutions do still collect specimens from the wild for valuable scientific research – with rigorous ethical considerations as to whether lethal means are necessary – if the object is on display the answer is almost certainly no. Museums today do not kill animals with the sole purpose of putting them on display.

However the vast majority of natural history museum specimens *were* killed at some point in the past. I understand why many parents choose to tell their children that all the animals they see in the museum were either found dead or died in a zoo (we hear this a lot), but the reality is that this isn't true. If you look hard enough you can often see the bullet holes to prove it.

Attitudes have obviously largely changed towards the acceptability of killing animals for any given purpose, particularly when it comes to vertebrates. However, as much as we may find it distasteful today, I would argue that we should all feel some degree of gratitude for the legacy the collectors of the last two centuries or so have left us in our museums. Without them, we wouldn't have the fabulous resources that provide so much inspiration, enjoyment and contributions to wellbeing, and act as scientific and social historical repositories that are critical to modern society.

There are different kinds of museums with different motivations, so it's a bit difficult to draw broad generalisations. Some museum collections were built primarily for public enjoyment, and they are often associated with local government bodies which means that their collections might largely relate to local matters. They may focus on local species or local collectors; though some will have a smaller 'global' collection to provide a window onto the wider world.

Others were amassed primarily for research. They can have giant collections running into the tens of millions of objects, with examples from species all over the world represented by many specimens of each. Because of the scale of the collection, a relatively tiny proportion will be displayed in their public galleries. These museums are often nationally administered organisations (like the Natural History Museum, London), or attached to large universities.

Others still will have their origins in the collection of one individual, who could have had any number of personal ambitions or specific areas of interest. Very often they involved collecting for a perceived public good, and they bequeathed them to public bodies upon their death. And many museums will have a combination of these histories – and others – having consolidated many different kinds of collections over the years under one roof.

Historically, many people employed directly by large museums did spend time away in the field personally collecting and preparing specimens to bring back, however it was more common to acquire them from others. The larger institutions would have led or commissioned expeditions specifically with the motivation to collect, while others would have relied on individuals working as professional collectors deployed around the world.

The logistics of visiting exotic locations means that some places were easier to arrange transport to than others, and there may also have been some political motivation to increase knowledge of a particular region. Knowledge of a country's natural history equates to knowledge of the potential resources – be they animal, vegetable or mineral – that could be exploited there. For these reasons, collections are often extremely biased by diplomatic relationships between nations. In the UK, it is easy to observe the bias of the former British Empire in what we have in our museums, and that is true of any country with a similar history. Collections of Australian species in British museums dwarf what we hold from China, for example.

When we see a beautifully prepared specimen in a museum display, it can be easy to miss the stories of the people who got it there. Unfortunately, in many cases these aren't recorded beyond the place and date the animal was shot, netted, trapped or clubbed. Many collectors worked through agents based back in Europe, sending them their specimens to sell on to museums. The existence of such middlemen likely made it more likely that the personal stories were lost or left in the background.

Only if a collector gained fame in other ways are we likely to have heard their tales. For example Alfred Russel Wallace – the co-discoverer of evolution by natural selection – earned his living by collecting exotic specimens to sell, chiefly in the Amazon and Indonesia. Museums around the world have Wallace specimens – particularly birds and butterflies – and they've come to cherish them because of the impact of Wallace's science. In these ways objects which were collected for science can take on a value for their role in our social history.

88 James Bond's Hutia

Type Specimens: Defining A Species

How do new species get added to the roll call of known animal diversity? By and large, species are the units of animal taxonomy. If a species is very closely related to other species, we put them together into a group called a genus (plural: genera). We put closely related genera into families; families into orders and orders into classes; and so on. Classifying species into groups like this helps us to make sense of the natural world, and allows us to speak in generalisations about similar species. This is essential given that there are untold millions of different kinds of animal. More than 15,000 new species of animal are named every year.

We use scientific names as a guard against the vagaries of language. For example, 'aardvark' may be the English common name in use today for that species (derived from the Afrikaans words for earth and pig), but they have many other names in many other languages. So we all know

This is the skull from the type specimen of James Bond's hutia – it is the physical definition of the sub-species. The safe-guarding of type material in museums allows for the long-term re-evaluation of the data that were originally used to describe the animal.

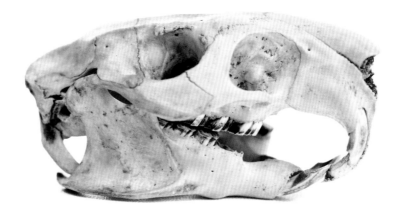

that we are talking about the same thing, scientific names are universal, and they are based on a system developed by the Swedish botanist Carl Linnaeus, 'the father of taxonomy'. The scientific name of the aardvark is *Orycteropus afer*. As it happens, aardvarks don't have any close living relatives, so they are the only species in the genus *Orycteropus*.

When a new species is described today (things were not always so rigorous in the past), if it is thought to be similar to existing species it is assigned to an existing genus. If not, it is given a new one. And then it is given a new species name. The combination of genus and species has to be unique – no two species can have the same scientific name (while 'robin' may refer to a number of species in different parts of the world, *Erithacus rubecula* is the unique name of the kind I get in my English garden). Alongside that, a comprehensive description is provided of the species' physical characteristics, and outlining clearly how it differs from its relatives. The level of detail can be extreme – species can be separated on the basis of the pattern of tiny bumps on their teeth, relative lengths of the sutures between neighbouring bones, numbers of scales, or any number of minute differences. Some differences are invisible, and two species may be separated by genetic differences alone. The name and the description are published in a scientific journal, after having been scrutinised and reviewed by experts.

This combination of a universal scientific name and a detailed description is intended to stop a new species from being described that is in fact the same as something that already has a name. However, it is not enough. Written descriptions could change in their interpretation over time (the meaning of words can change slightly), or new discoveries can be made that cause us to revisit the information that was originally used to describe a given species. New technologies arise that allow us to investigate species in different ways (Victorian scientists would have never foreseen what we can do with genetics, for example). The safeguard against this is the type specimen.

Type specimens are essentially the physical definition of a species: when the scientist describes a new species, they declare *an individual museum specimen* – called the holotype – (or occasionally a small set of specimens, called the syntypes, which can allow for individuals of different sexes and developmental stages to be described) with which the name is forever associated. The holotype of the aardvark, for example, is stored at the Natural History Museum, London. However language changes, or taxonomic groups get shuffled, whenever anyone sees an animal and says 'that is an aardvark', the true meaning is: 'that is the same kind of animal as the aardvark holotype'.

Type specimens are tangible objects which allow scientists to revisit the exact same data which were used to originally describe a species, sometimes hundreds of years later. This is a critical tenet of science – the same results should be achievable over and over again.

The specimen pictured on p. 279 is the holotype for a sub-species of Hispaniolan hutia (sub-species have type specimens too), called James Bond's hutia. Hutias are cat-sized, mostly tree-living, endangered rodents related to guinea pigs, restricted to the islands of the Caribbean. This specimen was found at the bottom of a tree along with a juvenile, apparently having fallen to their death. In 2015 James Bond's hutia was described, and was considered different enough from the other Hispaniolan hutias to be described as a new sub-species.

It is named after a Caribbean ornithologist, who was also the inspiration for the famous fictional spy. The hutia's discoverer Sam Turvey and I had a lot of fun discussing potential headlines for the press release: *You Only Live Mice*; *Live and Let Not Die*; and *From Hispaniola with Love*.

Depositing the physical remains of a new species as a type specimen in a museum is the gold standard of taxonomic practice, but there are a few exceptions. In recent years a number of primates have been claimed as new species using photos and live animals alone, with ethics and empathy used as the argument against killing for science. These drew significant challenges and are by no means universally accepted.

Less controversially in 2015 another species was described – this time of a bee-mimicking fly – which did not have a type specimen. The live animal escaped after it had been extensively photographed, and so it couldn't be added to a permanent collection as a type specimen. Instead, highly detailed photos stands in for the type – this was a world first. The authors suggest that species being described without physical type specimens will become more common. This is not because collecting a small number of specimens is likely to impact the population of an endangered species (particularly of invertebrates), but because regulations around the collection and export of specimens are becoming increasingly prohibitive to 'on-the-spot collecting'.

However, in the same publication the authors also make a strong case for using type specimens:

> Not only do they allow for consideration of a full suite of characters including internal morphology, microscopic and genetic characters, they preserve data for future access with future technologies and future questions. Specimen collections are our greatest treasure trove of biodiversity information and continued collection development must remain a priority.*

---

*   Marshall, S.A. and Evenhuis, N.L. (2015), 'New species without dead bodies: a case for photo-based descriptions, illustrated by a striking new species of Marleyimyia Hesse (Diptera, Bombyliidae) from South Africa', *ZooKeys*, vol. 525, pp. 117–127

89 Short-Beaked Echidna

MOUNTED TAXIDERMY

Man-Made Animals: The Wrong-Footed Echidna

When people think of natural history museums, among the first images to be conjured (perhaps after the dinosaurs) are of taxidermy specimens. The word taxidermy means 'arranging the skin' in Greek, but these specimens are very commonly described as 'stuffed'. This tends to irk professional taxidermists and some curators, partly because it isn't technically accurate, but also because such language dismisses the skilful craft that is required to create such specimens.

The hind feet of this echidna have been twisted forwards in order to meet the taxidermist's incorrect assumption that this is the way animal feet should point.

Rather than take an empty skin and stuff it full of packing material, taxidermy typically involves a rigid structure – called a form – being prepared in the precise shape of the animal's carcass, and the skin is then artfully arranged around it. These are often built by binding padding around a wire or wooden 'skeleton'. To keep the shape of the head and the appearance of the teeth accurate, many taxidermists find it is simplest just to reinsert the cleaned skull into the head skin, but otherwise the original 'insides' of the creature are all replaced by packing and supports. The specimen (and the form inside it) are mounted to take on a deliberate pose, to create the illusion or suggestion that the animal were still alive.

The pose they take on is at the whim of the taxidermist (or their client) – it could be sitting, standing, leaping, reaching, clawing or playing cards. There are several other kinds of specimens in museums that also seek to preserve the skin of the animal, but they vary in the extent to which they are intended to appear lifelike. Freeze-drying the whole carcass – insides and all – was pioneered as a display technique by Reg Harris, who was the Grant Museum of Zoology's curator in the 1940s and 1950s. This requires treatments that remove all the moisture from a specimen (so that it doesn't rot) involving freezing and desiccation in a vacuum chamber.

A similar result is achieved by chemically impregnating preserved specimens with wax or plastic. These animals tend to look as if they have recently died as the preparators don't always adorn them with false eyes, as they do taxidermy, so they clearly look dead.

I find taxidermy most interesting as it is both real and inauthentic. Pictured here is a taxidermy echidna. Echidnas are related to platypuses – together they form the only living egg-laying mammals, all found in Australia and New Guinea. The specimen here is *made of* echidna, but it isn't actually an echidna – it is its remodelled skin.

From speaking to museum visitors, it is clear that many people think that taxidermy in museums is in fact an animal that simply died and we have put it on a shelf. They are not aware that its insides are missing, and a great deal of work has gone in to making it look the way it does. Taxidermy is unquestionably man-made. This means that it is subject to human biases and misunderstandings.

The quality of taxidermy obviously varies in the same way as any creative output is dependent on the ability of the person creating it. However 'high quality' taxidermy that recreates a credibly life-like animal can also be deeply incorrect: it may look like a believably real animal, but maybe not the one that its skin belonged to. This can be because of the way that exotic taxidermy historically came into museums.

Imagine a group of explorers sailing on a voyage of discovery to unknown lands, or European colonists establishing settlements on the far side of the world. They would have been under instruction to collect animal specimens to send back home. However, they were not the ones that actually prepared the taxidermy. The likely scenario would have been that having killed an animal, they skinned it soon after death. At best, someone would have drawn a sketch of the whole creature, but they may have only written a brief description of it.

These notes would then be sent back to Europe along with the preserved skin and presented to the taxidermist. They were tasked with preparing a specimen in a life-like pose without actually knowing what the whole thing really should look like. This means that they had to make a lot of assumptions about what shape and stance the animal took, often based on hard-earned understanding of animal anatomy. Unfortunately their assumptions were not always correct.

Most mammals – obviously – have feet that point forwards. The toes are at the front of the hind foot. This is not the case for echidnas, which are adapted to digging vertically into the ground when threatened by quickly sweeping their backwards-pointing feet through the soil. This locks them into an immovable position with their vulnerable bellies protected underground, below the shield of defensive spines at the surface. Because the taxidermist did not know this, he (it was almost certainly a he) assumed that its toes should face the front, as they do in other animals. Unfortunately this involved twisting the legs through 180° until they ripped.

Such anatomical inaccuracies are rife in historic taxidermy. Most famously, perhaps, is the giant walrus at the Horniman Museum in south London. The extent to which it was overstuffed is a marvel in upholstery – the taxidermist didn't realise walruses are wrinkly, so kept filling the skin until it was smooth and tight.

But it's not just mistakes that demonstrate the inauthenticity of some taxidermy. By putting a snarling expression on a taxidermy tiger or fox – as was a common Victorian trend – the museum presents the animal as a ferocious beast. Such decisions may be a poor representation of the animals' temperament in life. Perhaps these were intended to suggest that the hunter had bravely bettered a dangerous animal.

Good taxidermy, however, is an art form. If they have the luxury of skinning the whole creature themselves – which is far more common in modern practice – taxidermists can take meticulous measurements of every muscle and bone as they dissect the carcass, and recreate them in their artificial form. Despite the extreme creative abilities required to make good taxidermy, the name of the 'artist' that prepared it has commonly been omitted from the object's history – perpetuating the myth that taxidermy is not man-made.

90 African Rock Python

MOUNTED SKELETON

Man-Made Animals: Skeleton Jigsaws

If we question the authenticity of taxidermy, are there other kinds of museum specimen that are subject to human bias? What about skeletons? It can be easy to look at an animal skeleton in a museum and miss the amount of work that has gone into building it.

In order to go from a dead animal carcass to a skeleton in a display case the bones first have to be retrieved, and then 'rebuilt' into the shape that they took within the body. There are a number of ways to remove all the skin, fat, meat, organs and other soft tissues from an animal, none of which are particularly pleasant.

Among the simplest is to bury the animal and let nature

After the soft tissue is removed, a huge amount of work goes in to rebuilding a skeleton like this rock python.

285

do most of the work – beetles, worms, woodlice and maggots eat the tissues, and eventually all that's left is the bones. Traditionally, this was a more common technique for larger animals, for the simple reason that the alternatives were impractical for whale-sized bones. Also depending on where you bury it there's no guarantee that you can find all the bones again, and that risk increases for smaller animals.

Many whale skeletons in museum collections would have come into being in this manner – sometimes with curators having to sacrifice parts of their own gardens to periodic interments of giant corpses. Indeed there was some talk that a forgotten whale may have been accidentally left in the grounds of London's Natural History Museum until the area in question was excavated in recent years for the building of a new wing.

For smaller animals, care has to be taken not to lose any of the bones. Rather than bury them in open ground, these could be put in buckets of wet sand, and again rely on wild decomposers to arrive and eat away the flesh. This is how this incredible African rock python skeleton came into being.

This massive snake arrived at the Grant Museum of Zoology as a whole specimen in the 1960s having died at London Zoo. Sadly the skin was in a poor condition and could not be preserved. Much of the work to prepare the skeleton had to be carried out on the flat roof of the museum.

More modern techniques of skeleton preparation include maceration (the process of de-fleshing by soaking) in baths of enzymes – basically biological washing powder – kept at the warm temperatures at which the enzymes best break-down the tissues. Often tanks of live insects called dermestid beetles are employed to eat away the flesh. In most cases a combination of techniques will be required to finally get to clean, dry bones ready for mounting.

Except with small animals, when the de-fleshing process can be halted at the precise time just to leave the cartilage that connects neighbouring bones, skeleton preparation results in a jumbled collection of loose bones which then need to be reconnected to recreate the shape of the animal in life. They can be fixed in place with glue, wire, and depending on the scale of the animal, huge metal rods, bolts or even scaffolding.

This is where human interpretation comes in. Although some skeletal joints are very tight, and only allow the bones to be arranged in one specific way, others are less precise and it's very easy to reconnect bones in the wrong order, or in the wrong place. It was only recently spotted that the rarest skeleton in the Grant Museum of Zoology and probably the world – that of an extinct zebra called a quagga – had its neck bones on upside down, for example. Many skeletons are arranged in slightly anatomically incorrect poses, with the limbs somewhat out of position, or the spine aligned too straight or too curved.

This isn't too surprising, if you think about it, as it would be quite easy to take a huge animal like an elephant skeleton and re-arrange it so that it stood on two legs like a person. This would be a very extreme example of the kinds of choice skeletal preparators have to make when working out how to rebuild a disarticulated skeleton.

At over 5m long, I find it very hard to comprehend the complexity of the work that the Grant Museum technicians undertook to rebuild this rock python skeleton. They are one of the five longest snakes in the world, and the largest in Africa. It is a jigsaw of thousands of pieces, when every one looks more or less the same: snake skeletons are comprised entirely of repeating ribs and vertebrae. I can only imagine that as the meat rotted off, they returned to the stinking sandpit on the museum's roof to number the bones as they were revealed, but this is just a guess.

An intriguing part of this story is that this animal came to the museum labelled as an anaconda, not a rock python – it was only recently that the photo of the whole carcass was looked at with a critical eye and it was re-identified. Misidentifications in museums are not at all unusual – people make mistakes – but the technicians who prepared it insist that they were told by London Zoo that it was an anaconda. So the question is, did it live its life at the zoo as a snake from the wetlands of South America, or the dry forests, savannas and semi-deserts of Africa?

The python carcass on the roof of the Grant Museum before it was prepared as a skeleton.

91 *Hypsilophodon*
CAST OF FOSSIL SKELETON

Dinosaurs In Museums

Dinosaurs are a defining feature of natural history museums. Standing next to the skeleton of a genuinely ginormous animal cannot fail to evoke some degree of awe. They were really, really, really big. Such encounters can cause museum-goers to ponder on the scales at which nature works, and perhaps sense some loss for the fact that evolution did not give us a chance to live alongside such land-giants and see them living in the wild.

When I give lectures about authenticity in museums, the students are usually extremely surprised that – aside from birds – an extraordinarily small percentage of dinosaurs on display in museums are actually real specimens. Nearly all of them are casts or replicas made of plaster.

If you think about it, this makes a lot of sense for a number of reasons. First, dinosaur skeletons, being fossils, are made of solid rock. Although there were a large number of smaller dinosaur species – including this one, *Hypsilophodon* from the Cretaceous of England – the 'best' museum dinosaurs are massive. This means that they would weigh an enormous amount. Mounting a skeleton of such a weight presents a number of mechanical difficulties in both building a scaffold that is strong enough to hold all the fossil bones safely and reversibly (museums don't want to make permanent changes to their specimens), and ensuring the floor of the gallery can take the weight.

Second, the odds of every bone from a large skeleton being fossilised and surviving from the time of death to current day is vanishingly small. It's possible that a complete large dinosaur skeleton has in fact never been found. Most specimens we see in museums are replicas of composite skeletons – different parts of the skeleton come from different individuals of the same species. Palaeontologists make casts of appropriately sized bones from a number of partial skeletons to make up a replica of a complete one. In most cases they will actually have to create some of the replica bones from scratch, based on inferences they make from neighbouring elements and related species, as they have never found every bone of some species. We don't actually know how many vertebrae, for instance, there should be in the spine of the giant herbivorous sauropods like *Diplodocus*.

One of only two real, large mounted dinosaur skeletons on display at London's Natural History Museum, for example, is the world's most complete stegosaur, which has only been in the galleries since 2014. It is 85 per cent complete, which is pretty darn good. The 'missing' bones have been replicated so she appears complete, and these bits are labelled so as not to deceive. Interestingly, in 2017 the same museum – rightly in my opinion – removed their most iconic specimen from display: 'Dippy' the *Diplodocus* was replaced by the skeleton of a blue whale. The reasoning was that the whale far better illustrates issues around biodiversity, and it is a real skeleton – unlike Dippy, which was a plaster cast of a composite. The sense of awe is not lost: blue whales are the biggest animals to have ever lived.

By making multiple casts of the composite skeletons, museums can then share the material and exhibit exactly the same replica skeleton in different parts of the world at the same time. This is undeniably a good way of allowing more people access to such incredible material.

While all of these explanations are perfectly reasonable, what really surprises the students in my lectures is that they weren't already aware of it. In order to better instil the sense of wonder, and to make the whole experience more authentic, museums make their plaster casts look like real fossils – they paint them brown and black. White plaster replicas just wouldn't have the same impact (as well as being a nightmare to keep clean). This enhances the visitor experience, but in striving for authenticity in a sense they become less honest.

There is nothing wrong with this, as long as the skeletons are labelled as replicas. Museums have a real responsibility to maintain the public's trust and not to mislead their visitors. Of course, museums do label their dinosaurs as casts (at least most of the time – there are definitely exceptions and I'd encourage you to keep an eye out), but the problem is that visitors don't read every detail on every label. Should museums be doing more to communicate that replica dinosaurs aren't 'real'?

92 Dodos

SUB-FOSSIL BONES

Dodos In Museums

Dodos and dinosaurs share a lot in common: they are both extinct, they are both firm museum favourites, and as I will discuss they are both – on the whole – only represented in museums by their skeletal remains, as composite skeletons made of the bones of many individuals. Where they differ, of course, is that humans actually encountered living dodos. Unfortunately.

A large collection of bones from several dodos, including pelvises, breast bones, leg bones, vertebrae and a lower jaw.

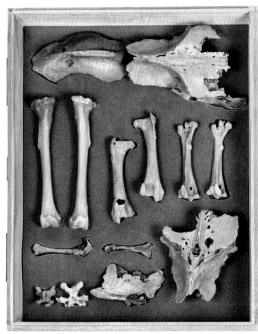
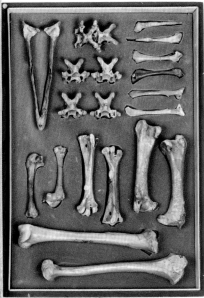

The dodo is a true icon of extinction. It was likely the first species that humans realised they themselves had completely exterminated. This is probably why, whether consciously or not, museum visitors are so keen to see them. Dodos lived on the island of Mauritius in the Indian Ocean, which was completely devoid of mammalian predators until Dutch and other European sailors arrived there in the early 1600s. As has been discussed, such predator-free conditions on islands often lead species to lose their defensive adaptations, both in their behaviours and their anatomy. Dodos – related to pigeons – over time on their isolated island had lost the power of flight, grown rather round, and began nesting on the ground. They also lost their wariness of other animals that might want to eat them.

Dodos were doing just fine by not wasting energy on defences on an island with few enemies, but that changed when people arrived. They not only hunted them (with great ease – they could be approached and clubbed), but also introduced animals that would eat them and their eggs: rats, cats, dogs and pigs. They were all dead within a century.

Despite the fact that they were considered to be of great interest to Europeans, there are few reliable depictions of what an actual live dodo looked like. The image that is most prevalent was conjured for *Alice's Adventures in Wonderland* by Lewis Carroll, who lived two centuries after their extinction. The contemporary drawings – mostly by sailors – are frustratingly inconsistent, and it's not always clear if they were first-hand witnesses.

Their physical remains were not much help either. Prior to 1865 dodos in museums were extraordinarily rare. At that time there were just remnants from a damaged skin in Oxford University Museum of Natural History; a foot in what is now the Natural History Museum, London (since lost); a skull in the Natural History Museum of Denmark; and a partial skull and leg bones in the National Museum, Prague. All these specimens, as you would expect for modern species, were obtained directly from living animals (that is to say, they were removed from the animals very near the time of their deaths).

By contrast, nearly all the dodo material in museums today – including the bones pictured on the previous page – was collected from animals that had been long dead. More than 99 per cent of all dodo specimens come from a single swamp – the Mare aux Songes.

In 1865 a teacher named George Clark, who had long been looking for evidence of dodos on Mauritius, successfully led a search of the swamp which turned out a few bones. Then by directing local labourers to wade in the marsh, feeling for the bones in the mud with their bare feet, he discovered the majority of bones currently known. These specimens were then sold or donated to museums. It is remarkable to think that nearly all

The world's only surviving soft tissue from a dodo, from the Oxford University Museum of Natural History.

of the existing specimens of such an iconic animal come from one tiny place, and were collected by one man.

Many museums have on display a 'stuffed dodo', but they are not real. They are made as illustrative reconstructions from pigeon, chicken and turkey feathers (among others).

In truth, the only known soft tissue of the dodo anywhere in the world is at the Oxford University Museum of Natural History – it is the leathery, largely naked skin from the head and foot. I have never been so excited to see a museum object as I was the day I was in the storerooms there and the dodo's cupboard was opened.

This specimen started life-after-death as a complete dodo skin, so it's slightly devastating to think that we could have a complete skin of the species if only circumstances were slightly different. The story of what happened to the Oxford dodo is extremely widespread in museum circles. What most people have been told about this skin is that in 1755 the museum's trustees decided that it had suffered too much damage from insect infestations to display, so they threw it on a fire. The head and foot were then rescued from the flames by a diligent member of staff. That is all that remains of the dodo today.

As dramatic as that sounds, it is in fact not true.

Where does this story come from then? The Latin word for inspection is very close to the word for fire. The records of the University were written in Latin, and this is how the fire was inserted into the dodo's history – the dodo was inspected in 1755, not burnt.

A number of museums hold dodo skeletal material collected by George Clark, and a few have decided to mount them as partial skeletons. Like many dinosaurs, however, these bones were not collected together and represent composites of many individuals.

93 The Micrarium

2,323 MICROSCOPE SLIDES

Museum Galleries Don't Represent Nature

It shouldn't be surprising to hear that I am a huge fan of museums. They inspire people to care about nature, they hold and preserve specimens in public trust, and they carry out critical research in taxonomy, ecology and conservation biology. However, as much as their galleries are often incredible and awe-inspiring, natural history museums don't represent the natural world very accurately.

I've mentioned a few of the ways in which biases can creep into the way museum objects are prepared, through human error or even politics, but there is a wider bias visible across most natural history galleries as a whole: there are a lot of big animals on display, and relatively few small ones.

Pictured here is the Micrarium at the Grant Museum of Zoology – a place for tiny things. It is believed that when we built it in 2012 it was the first of its kind, displaying 2,323 microscope slides and 252 lantern slides lining the walls on floor-to-ceiling light boxes and the effect is quite staggering. The slides mostly show whole small animals, or slices through whole small animals, a preparation technique which itself is amazing. Imagine taking a slice 1/10th of

The Micrarium at the Grant Museum of Zoology gives some attention to the tiny animals that make up nearly all of global diversity.

a millimetre thick through a fly, cutting through its antennae, its body, its head, its wings, its legs and the hairs on its head, all at once.

The Micrarium was conceived and built to right two wrongs. The first is that displays in natural history museums are deeply unrepresentative of nature. As has been mentioned, nearly all animals are invertebrates. 80 per cent of all described species are arthropods alone – as evolutionary biologist JBS Haldane [probably] said 'the Creator, if he exists, has an inordinate fondness for beetles', and yet I'd be surprised if 10 per cent of museum displays focussed on them. That's no surprise – dinosaurs and monkeys are a much easier sell than bugs and worms.

We can relate much more easily to animals like us with skeletons, babies, faces and two eyes. Invertebrates are so alien with larvae, suckers, segments, antennae and various kinds of exoskeletons that it can be very hard to 'relate' to corals, acorn worms, sponges and spiders. I know I'm guilty of a similar bias: only forty-two of the 100 objects in this book are invertebrates. The fact that vertebrate skeletons preserve so much easier than invertebrates doesn't help invertebrate representation either.

The second driver behind the Micrarium was to solve the issue of microscope slides in museums. We have around 20,000 microscope slides in the Grant Museum of Zoology – that's nearly a third of our entire collection. Yet not one was on display. Actually I did a survey of all the natural history collections in the country to see how this compared across the sector. This is what I found:

Lots of museums hold thousands of zoological microscope slides.
Few have any on display.
Those that do only display a few, or at most tens of slides.
Slide boxes are displayed as exhibition props: set-dressing to illustrate the office or work place of a scientist.
Some museums have put a few under a fixed microscope for visitors to look at, and some have tried displaying close up images of them.

In truth, microscope slides are really problematic for museums, and it's no wonder the museum sector has struggled to know what to do with them. For one, they aren't very useful anymore. The main reason we have so many is because microscopy used to be an integral part of biological study and research. Students gained significant zoological knowledge from the practice, and over time we built up a massive collection of teaching slides. At University College London, students would each borrow a set (the deposit was two pounds and two shillings), like a library book, and be referred to particular slides during their lectures.

The older slides in the Micrarium were bought by the likes of Robert Edmond Grant from slide dealers in the mid-nineteenth century for use in his teaching or research. Many of the slides, including serial sections of animals, embryology and different developmental stages, were prepared for scientific treatises on comparative anatomy and how different animals grow. The use of microscope slides in teaching continued until relatively recently, but has been made obsolete by better imaging techniques and a greater focus on genetics rather than morphology. In research, it is only being used at a fraction of previous rates, and historic slides are referenced only rarely.

Another issue museums have with slides is that they are tiny. How do you label them so that everyday museum visitors understand what they are looking at? Any label which conformed to the Disability Discrimination Act would be at least 4 times as big as the slide itself. That's assuming that the specimen on the slide is visible with the naked eye anyway (after all, the whole point of microscopy was to show things that weren't).

The Micrarium at the Grant Museum of Zoology was built as an experiment. We were trying something new with no real idea of how people would respond. Given their sheer number – meaning that attempts to label them individually would be impossible – the intention isn't for visitors to get specific insights into individual specimens or species, but to appreciate the sheer vastness of invertebrate diversity. So far, the experiment is working – it is extremely popular, and has been replicated in a number of other museums.

These specimens are exquisite works of art, and this installation is extremely aesthetic. It aims to inspire awe at both biological diversity and human technical skill in their preparation. Visitors don't have to be versed in worm taxonomy or the finer points of flea anatomy to appreciate what they are looking at.

Some of the microscope slides in the Micrarium.

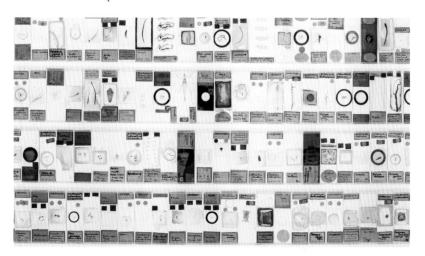

94 Fruit Flies

PRESERVED SPECIMENS

Model Organisms And The Decline Of Zoology

The ways in which animals have been studied and taught about have diversified and grown over the centuries since the Enlightenment. At various points in history the particular focus of the kinds of information that were in vogue for zoological investigation have shifted. Classification, anatomy, comparative anatomy, evolutionary biology, animal behaviour, ecology and conservation biology – among others – have inspired the dominant research questions at different points in time. At any one time, one field was often emphasised while the others received less interest.

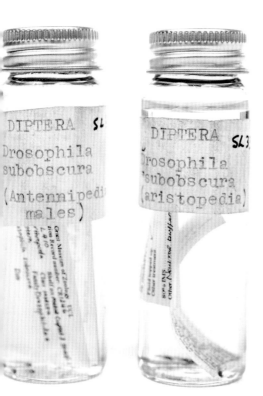

Museum collections, particularly those associated with universities, reflect these changing trends because the objects that get added to the collections often came from the researchers and lecturers of the time – we can see what was happening in zoology by looking at what specimens were being collected.

By looking at university collections from the 1960s to early 2000s, we can conclude that these were dark days for the teaching of zoology. What students get taught obviously depends on what their lecturers want to teach them.

Commonly called 'fruit flies', but actually members of the vinegar fly family, these insects have been used for countless studies into genetic systems.

Trends become perpetuated as the students' view of zoology is narrowed depending what their teachers focus on, and what elements of the discipline they decide are less important. In many places during this period the study of whole animals – what I like to call 'proper zoology' – suffered a serious decline.

At this time the genetic and biochemical advances were very much at the forefront of biology and the curricula of many universities focussed on these emerging disciplines to teach their students. There is no doubt that these are essential branches of the study of life sciences, but by focussing on them the students missed out on the areas of study that had a much longer history such as taxonomy and evolutionary biology. These subjects were considered old fashioned and dusty and were afforded much less time – if any – on biology courses. These attitudes were then transferred to the museum collections with which they were associated.

Biology departments in universities across the country sought to make space for the more 'modern' research areas by clearing out their museum collections – if they weren't using their specimens for teaching anymore, how could they justify keeping them? As shocking as it sounds, centuries' old objects, painstakingly collected, preserved, documented and displayed were confined to the skip.

As this was happening, animals were being taught about in a different way. The complexities of biodiversity were reduced to focussing study on a handful of species, known as model organisms. Animals such as fruit flies, guinea pigs, rats, mice, dogs and monkeys were used as 'models' to understand the whole world. The idea is that by investigating the biology of a few species, students could generalise about how *all* animals functioned. At the Grant Museum of Zoology, for example, there are boxes that contain hundreds of skeletons of frog and rabbit legs, pigeon wing bones and guinea pig skulls.

Model organisms have been a critical part of medical sciences for centuries. The use of lab mice and the like have allowed drugs and treatments to be thoroughly tested before they were applied to human subjects. Innumerable everyday medicines, transplant techniques and vaccines have only been made safe by investigating how non-human animals with comparable biological systems respond to them.

Much of what we know about how our genes are inherited and function, how embryos develop and how our brains work have come from the study of model organisms – they are a critical element in biology, and we would be at a loss without them. Although there are serious ethical considerations about the use of animals in experimentation, they have saved and improved countless human lives across the world. However, that shouldn't have meant that the way biology was researched was reflected in our classrooms. 'Old fashioned' zoology was sidelined.

In biology it is often said that whatever scale your research operates at, you need to understand the scales that are smaller than it, as each level of complexity underpins the next. You cannot understand large-scale biology, like ecosystems and ecology or evolutionary biology and classification without understanding small-scale biology, like anatomy, genetics and molecular biology. To study whole animals you have to study genes and molecules. But to study genes and molecules, you don't need to study whole animals.

For these reasons, as genetics and model organisms were pushed to the fore, proper zoology suffered a downfall. Irreplaceable specimens in museums were senselessly destroyed under the guise of progress. The science and the students suffered. I am not exaggerating when I say that biology students were graduating that could not tell the difference between the skull of a sheep and a dog.

Fortunately, it seems that in the last decade or so, universities have realised that you cannot teach a student what a wombat looks like by studying its genome. The ludicrously obvious conclusion appears to have been drawn that a balance is needed in how biology is taught: a bit of genetics and a bit of whole animal biology. Unfortunately, in many universities which short-sightedly disbanded their zoology departments and their collections of specimens, the students there no longer have the best resource for teaching zoology – museum specimens.

Frogs were widespread model organisms for the teaching of general principles in biology. These legs most likely resulted from student classes.

95 Grass Snake

PRESERVED ALIZARIN STAINED SPECIMEN

Keeping It Wet: Preserving In Jars

There is a marked difference in many museums when you compare what is behind the scenes and what is in the gallery. Most natural history museums only display a fraction of one percent of their collections, which is not surprising as the larger ones run into the millions of specimens.

In the storerooms, cupboards and drawers live the specimens that were not intended for public consumption. Many objects were prepared with display in mind: taxidermy, as we have discussed, is a painstaking process that wouldn't be undertaken if it weren't going to be shown. There are much simpler ways of preserving a tanned skin: storing them flat is common for larger mammals, while birds and other smaller creatures are usually packed with a little filler and stored as a roughly cylindrical 'study skin' in a drawer. Likewise a

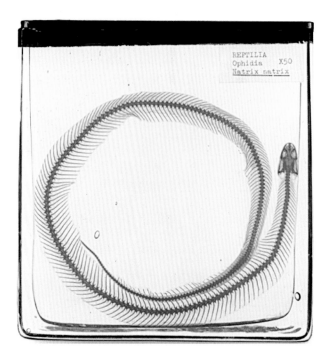

REPTILIA
Ophidia   X50
Natrix natrix

Alizarin staining makes the soft tissues appear colourless and the skeleton is stained red. This allows the position of the bones in the body to be observed.

posed skeleton is far more costly to mount and takes up far more room than a box of loose bones.

These less flashy objects are amassed – often outnumbering their counterparts in the show cases – because display for the public's benefit is not the only purpose of the museum. Collections in museum storerooms are vital for zoological and historical research, allowing the same material to be studied by people living hundreds of years apart.

I'll explore the decisions behind what does go on display in the next chapter, but one type of object that is found in the gallery far less commonly than taxidermy and skeletons is the fluid-preserved specimen: animals or bits of animals stored in jars. They are also called wet specimens, spirit collections or 'pickles', which can rankle some museum workers as they feel it diminishes the skill of their profession in the way that calling taxidermy 'stuffed' does.

While some animal groups, particularly aquatic species and small reptiles, lend themselves to display as wet specimens, it is relatively rare to find mammals and birds in jars in the gallery: these live behind the scenes.

Wet specimens preserve all the soft tissues – including even the stomach contents – and so are an incredibly valuable tool for science. They are far less subject to the modifications and losses that take place with some of the other museum techniques, and allow us to investigate the internal anatomy.

In order to keep an animal in suspended animation in a jar the process of decomposition has to be halted, ideally as soon as possible after death. Most wet specimens are first 'fixed' with a fixative solution, which stops the carcass from breaking down at a cellular level – normally a dead animal decays when its proteins break down and its cells liquefy. Fixatives stop this, forming chemical bonds to coagulate the cell contents.

The most common fixative in museums is called formalin, which is a rather unpleasant solution of the highly toxic gas formaldehyde. There are many hazards to working with museum collections, and this is one of them. (Another is that until relatively recently it was common to coat taxidermy and skins in arsenic or other poisons to stop them from being eaten by insects.) To work, the fixative needs to penetrate all through the specimen, so it is either injected or the carcass is cut open and bathed in it.

After being fixed the specimen is then stored permanently in a preservative. Formalin can be used for this too, but it's far more common to use an alcohol. Indeed, sailors on voyages of discovery around the world would often find they had to share the rum with the naturalist on board, who would be using it to store valuable animal specimens (rendering it unfit to drink, before you ask).

However, drinking spirits are not strong enough to preserve a specimen for the long term. Most museum specimens today are stored in ethanol – the same alcohol found in booze. Trends vary around the world and through time, but the UK today, for example, favours a solution called Industrial Methylated Spirits, which is typically 70 per cent ethanol with a small amount of methanol in it. The methanol should just about put off anyone tempted to drink it.

Because alcohol is so volatile, and it's very difficult to make a perfect seal on a jar, the level of preservative in fluid specimens can drop over time, but it can be topped up relatively easily. A well prepared specimen can last for centuries in this way.

Unfortunately, the fixation process can render it very difficult to use preserved specimens for genetic studies (something our predecessors could not have predicted), as the formalin forms crosslinks with the DNA (and if it doesn't DNA degrades quite quickly anyway). Nevertheless useful DNA samples have been retrieved from such specimens.

The chemical properties of organic tissues can be manipulated by different preservation techniques to show some pretty astounding things. The grass snake specimen displayed on the previous page has gone through the process of alizarin staining. This renders the soft tissues clear and colourless, and the bone is stained red. In this way we can see exactly how the skeleton is arranged in the body. Such treatments avoid the risk of human error in reconstructing a cleaned skeleton. They are also rather beautiful.

By keeping the whole body of animals like this tarsier in fluid, the internal anatomy is preserved for long-term study.

96 Domestic Cat

PRESERVED BISECTED PREGNANT SPECIMEN

What Goes On Display?

We have already discussed a number of ways in which museums are not the objective purveyors of 'true' knowledge that people might hold them up to be. This is reflected in the ways that some specimens are inaccurately prepared; the biases that underlie what animals get collected and where they get collected from; and what decisions – subconscious or otherwise – are made about what parts of the collection should be put on public display. It is this last topic that I'd like to explore further here.

Are some specimens too gruesome for museum visitors to see on display? Familiar (and cute) species like this cat can evoke negative responses.

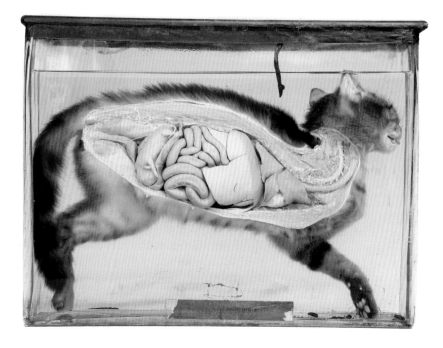

Why is it that fluid-preserved specimens are displayed less regularly (at least for some animal groups including mammals and birds) than taxidermy or skeletons, even though they are potentially far more accurate? If we compare wet specimens with taxidermy, one answer presumably lies in the sense that taxidermy is intended to convey the deceit that the animal is still alive (as well as the fact that jars of preservative for larger animals are impractically heavy and dangerous). But what about skeletons? There are a number of factors at work, as what we can learn from a skeleton is very different from what we can learn from a whole preserved animal, but I suggest that it is far easier to dispassionately view the skeleton as a cold, objective 'object'. An animal in a jar, however, was obviously a once-living creature, and now it is not. I suspect museums are shying away from displaying animals in jars as visitors find them more disturbing than the alternatives.

If this is true, it is interesting as it implies the average visitor is happy to be lulled into the false impression that museums did not historically kill animals. I know from countless conversations with museum visitors that they consider fluid-preserved specimens more 'cruel'. And none more so than a specimen that has been dissected.

This isn't surprising considering such specimens are overtly visceral, unlike a skeleton, which is clean, and taxidermy, which can be cute. Needless to say it's irrational, as in the majority of cases the animal would have been killed no matter which preservation technique is used.

I have encountered few objects that cause visitors to have such a strong negative response than this bisected cat (even though few recognise that it's pregnant), and this is interesting too. They seem more concerned about this cat than when they are confronted with the preserved remains of endangered, exotic creatures. The human connection with this species is so strong – even though it is one of the most effective agents of eco-logical destruction on the planet – that many people find it challenging to see them preserved in a museum. (This remains the case when we have displayed the cat the other way around, keeping the dissected side out of sight.) Yet a species on the edge of extinction similarly displayed does not evoke the same concerns. As a result – I believe – museums shy away from displaying such material, even though it is common in many storerooms (after all, cats were a significant model organism for biological research).

There are other reasons to think that museum curators modify their displays to cater to the sensibilities of their visitors. The majority of mammal species, for example, have a bone in their penis. It's called a baculum. A handy mnemonic to remember which groups have them is PRICK: Primates (but not humans), Rodents, Insectivores (an outdated name for shrews, moles and hedgehogs), Carnivorans and [K]Chiroptera (the scientific name for bats – I admit this is a bit of a cheat). Despite the prevalence of skeletons of these animals in museum displays, it is extraordinarily rare to see one with its penis bone attached. One reason for this is the presumed prudishness of the curators, who would remove the baculum before putting them on display (another is that they are easy to lose when de-fleshing a skeleton).

Being sensitive to visitors' objections is only one way that human bias creeps into what curators have chosen to put on display. Another is revealed if we think about the sex ratio of animal specimens in museum galleries: the males are thoroughly over-represented, and the females are sidelined.

Rebecca Machin published a case study in 2008 of a typical natural history gallery and found that only 29 per cent of the mammals, and 34 per cent of the birds were female. To some extent this can be explained by the fact that hunters and collectors were and are probably more inclined to acquire – and been seen to overcome – animals with big horns, antlers, tusks or showy plumage, which typically is the male of the species. But can this display bias be excused? It is a misrepresentation of nature.

Furthermore, Machin found that if male and female specimens of the same species were displayed together, the males were typically positioned in a domineering pose over the female, or just simply higher than her on the shelf. This was irrespective of biological realities.

Looking at the ways in which the specimens had been interpreted – even in labels that have been written very recently – she found that the role of the female animal was typically described as a mother, while the male came across as the hunter or at least had a broader role unrelated to parenting.

We have to wonder what messages this might give museum visitors about the role of the female. Even in the apparently scientific environments of natural history museums, animal displays appear to reflect historic social norms of people, in the absence of any scientific fact. One has to assume that these prejudices were accidental, but it does cast the objectivity of museum displays in a different light.

97 Copse Snail

BLASCHKA GLASS MODEL

Preserving The Un-Preservable: Blaschka Glass Models

Many of the Victorians who put together our museums were effectively seeking to assemble comprehensive compendiums of the natural world. They sought animals from around the world and from throughout geological time to create a microcosm of human knowledge in their galleries and cabinets. These were used to instruct, inspire and impress.

I have discussed a number of biases that are detectable in natural history museum displays, resulting from misconceptions or misunderstandings about animal anatomy, prevailing socio-political norms around gender and colonialism (among others), and good old fashioned prudishness and taste. In the majority of cases the museum curators of the time would have been unaware that their personal prejudices were creeping into the way they were displaying the objects.

However there was one museum bias that they did seek to rectify – one that could potentially stand in the way of museums being able to make comprehensive displays of animal life: the issue that some animals could not be easily preserved.

In the last chapter I proposed a bias against displaying preserved specimens in museum galleries, but this was largely relating to mammals and birds – the groups to which we most closely relate. Invertebrates – animals which have traditionally not enjoyed the same level of concern regarding cruelty – are often displayed in jars of fluid, particularly marine species. This makes a lot of sense as they can't be stuffed, and they have no bones – the only alternatives are to store the dry remains of arthropod exoskeletons, stony corals, echinoderm tests (the name for the skeletal structures of sea urchins and starfish) and mollusc shells.

The problem facing nineteenth-century curators, however, was that they couldn't successfully pickle them. Many of the soft-bodied invertebrates such as jellyfish, corals, anemones, molluscs and siphonophores would lose all their colour and become disfigured when they were bathed in fixatives or preservatives. The chemical pressures resulting from these treatments would result in all the water – which makes up a huge proportion of their bodies – rapidly diffusing out. This would leave them shrivelled, broken and unrecognisable – completely useless for museum displays.

In a museum that sought to be comprehensive, how could they operate without these major animal groups? They would leave significant holes in their taxonomic displays. In the collections that were also teaching students, this would mean sizeable portions of the animal kingdom would have been omitted from their courses.

This snail has been perfectly modelled out of glass by the Blaschkas. Such specimens have become world-famous works of art.

The solution was to replace them with models.

A father and son team – Leopold and Rudolf Blaschka – have become world famous for having produced some of the most intricate, anatomically accurate and beautiful models of marine invertebrates out of glass. A Czech family working out of Dresden, Germany, in the second half of the nineteenth century, the Blaschkas originally made their name as jewellers, but they then ran a successful business supplying museums and universities across the western world with their incredible, lifelike aquatic recreations. They were later commissioned exclusively by Harvard University to make anatomical glass models of flowers.

By blowing, moulding, painting and staining glass to perfectly recreate the soft-bodied animals of the sea, museums could fill the gaps where their preservation techniques had failed. The idea of a natural history museum with no jellyfish, anemones and octopuses seems deeply problematic, and the Blaschkas solved this problem.

While preservation techniques have since improved for preserving some of these animals in fluid (but not their colours), the methods of the Blaschkas have largely been lost – no one has been able to recreate their work to the same degree of accuracy using the tools they had at the time. Only in very recent years are researchers beginning to understand the processes that went into shaping these creatures out of glass, by CT-scanning and x-raying the models.

These objects are found in a number of museums across the world and have now become real treasures of the institutions that care for them. Unquestionably they are exquisite works of art, and yet they do not belong in art collections. Instead, they are typically displayed alongside 'real' zoological specimens from the groups they represent. The Blaschka models are so perfect that in a natural history display it's often unclear which objects were once living and which are made of glass.

My favourite thing about them is that their history as taxonomic gap-fillers means that they are displayed like any other natural history object: labelled with the same labels and often put in the same jars, only without the alcohol.

Their fragility terrifies me every time I have to handle one. The glass is extraordinarily thin, particularly on their finger-like projections and tentacles. What I can never quite believe is that they were ordered as standard out of a catalogue, and then shipped in crates from Germany to all corners of the western world. The very idea of a horse-drawn carriage containing these incredibly delicate sculptures rumbling along the cobbles of Victorian London fills me with dread.

NATURAL HISTORY IN 100 OBJECTS

98 Red Panda

So When Is Natural History Art?

Much has been written about the cross-over between art and natural history, particularly when traditional scientific museum practices are replicated in art. I described the Blaschka glass models as works of art in the last chapter, but they are also natural history specimens. The craft and skill that goes into preparing a huge number of museum specimens is also unquestionably artful, but is it art? I don't want to get into the stagnant topic of 'what is art?', but it is interesting to consider what makes some objects 'art' and some objects 'science', when we might struggle to tell the difference if their context were stripped away. In some cases, more or less the same thing is created in the name of art or science. The obvious answers relate to the intentions of the artist and the interpretations of the viewer.

In recent decades, one of the world's most prominent artists who has made use of natural history museum techniques is Damien Hirst, who regularly includes dead animal specimens in his pieces. One of his more celebrated pieces is *Mother and Child Divided*, 1993. It comprises real dead specimens of a cow and a calf each cut in half lengthways, and preserved in four huge tanks of formaldehyde – one tank per half of animal. They are arranged to allow people to walk between the animals' two halves and inspect the insides. Ooh! Art. Obviously. It's amazing, full of imagery, suggestion and meaning.

Bisected specimens are common in natural history museums. Consider the bisected red panda head pictured overleaf. Aha! Science. Quite clearly. Beautiful, intriguing, inviting closer investigation. It was prepared as part of a large collection of bisected animal heads by Sir Victor Negus at the Ferens Institute of Otolaryngology – they allow the study and comparison of structures in the nose and throat across a huge range of animal groups. Yep. Science, for sure.

It is very likely that Hirst has seen these specimens. The technique is the same, they look the same. Are they the same? I'm unsure.

Having walked between the two halves of Hirst's cow I, like everyone else, inspected recognisable aspects of the anatomy. 'Look at the stomach!' was the paraphrased response of about half the people I eaves-dropped on in the gallery. Isn't that science?

If you have ever milked a cow you will know that it feels like a thin rubber bag filled with liquid, which you pinch a bit off of and squeeze

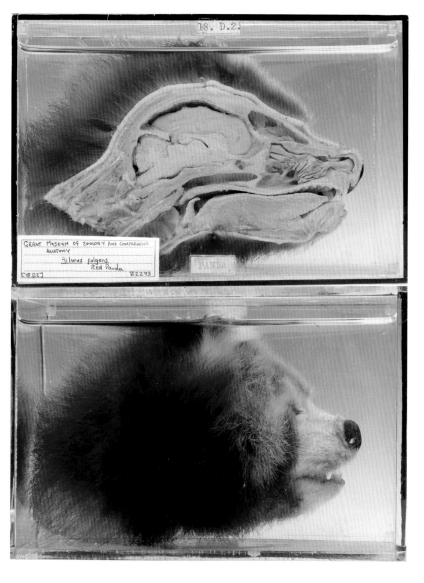

out the contents. It doesn't feel like a human breast. Looking at *Mother and Child Divided* I was, for the first time, given the opportunity to see an udder cut in half, and try to work out the physical reality of my milking sensation – which sounds to me like scientific investigation.

At first glance, then, the piece is art – the title is a nice, suggestive pun, but then people start to investigate it closely and look at its anatomy. Surely by this point the artist's intention is forgotten and the viewer's interpretation turns it into a natural history museum object?

When I visit art exhibitions which incorporate natural history objects, like Hirst does, I end up chiding myself for failing to switch off my museum-visitor setting. I search the lists of materials they put on the labels for species names (common ones, not scientific ones) of the animals in the artworks. I know that isn't the point of the art, but it's force of habit.

Perhaps the most famous split-personality artwork/specimen is Hirst's *The Physical Impossibility of Death in the Mind of Someone Living*, 1991 – a fluid-preserved shark in a huge Perspex container of a size that is extremely rarely seen in natural history museums (they are extraordinarily expensive, not to mention the risks of storing that much flammable and toxic liquid in one place). When I saw it in person, I heard ten or twelve other visitors asking the same question I wanted answering: what kind of shark is it? It was a tiger shark, but on the list of materials on the label it just said 'shark'.

Beyond its title, this specimen should/could be in a natural history museum, it's just the fact that we can't typically afford tanks that size which makes it art.

Natural history museums include some incredibly beautiful, exquisitely prepared specimens that can instil a sense of awe and wonder in the viewer, just as art can. But on the whole they are considered differently when seen in the different contexts of the art gallery or natural history museum. The titles and context provided by the gallery – and the intention of both the artist/preparator and the viewer – are decidedly different. In the natural history museum we don't often consider the 'creator' of the specimens, while the name and identity of the artist is typically the chief concern in the gallery.

So to massively oversimplify, and to unfairly compare Damien Hirst's life's work with countless nameless specimen preparators, one possible conclusion is that science has the ability to do everything art does, but also answer the questions that we just can't stop ourselves asking. Would it take anything away from Hirst's masterpiece to stick the word 'tiger' in front of the word 'shark' on the label? No. No it wouldn't.

Of course, it's perfectly possible that I'm missing the point.

99 Human

Putting Human Remains On Display
– People As Animals

This is a book about the animal kingdom, told through the stories of museum objects. There is one kind of museum object that is different to all the others, at least in the eyes of the law. Regulations apply to the ethical and legal collection, acquisition, trade and exploitation of animal specimens generally, but one species of animal has legislation all of its own: humans.

In the UK, the Human Tissues Act (2004) controls how human bodies, organs and tissues are used. Different licenses are required to store, teach with, research or display human specimens that are less than 100 years old. Assuming the licence requirements are satisfied, there are many ways in which a human might find themselves in a museum after they

A human specimen among brains of other animals in a comparative anatomy display.

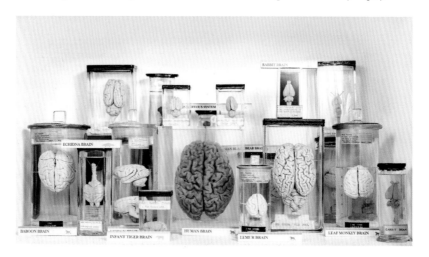

died. It's probably true to say that they are the species we are most likely to encounter as an object in a museum.

In archaeological galleries the human is often there to represent their culture at a specific point in time – often appearing in the form of burial remains. The museum's interpretative information about the remains might explain what the burial situation and scientific analysis of the bones tell us about their culture. They may also find that details of their own specific lives are included in labels – who might this person have been?

In anthropological galleries a human may find themselves simply as part of an artefact made by someone else. For example in a tooth necklace or a skull-cap bowl. In this way they may not be there to represent themselves individually or even their own group – they are reduced to an object made by somebody else, representing somebody else's behaviour.

In social history museums human remains may be there to physically represent the very individual that they once were. The label might read 'This is Queen Nancy ...' The object that Nancy has become is there not as a representation, but as the real, specific thing – Nancy herself. The labels would likely tell you about Nancy's life. She continues to have an identity, and her personality is the focus of the story.

In medical collections human remains are often displayed as a representation of a specific condition – a lung displaying cancer, for example. Any specifics of the person's life (beyond factors linked to the medical condition in question) are likely omitted from the labels.

In natural history collections, however, human remains probably have to work harder than they do anywhere else, as they are normally there to represent their entire species. Just like every other object we display – this is a platypus; this is a turkey; this is a human.

So to recap:

Archaeology – remains may represent populations in a given time and place.
Anthropology – remains may simply be part of an object made by someone else.
Social history – remains may represent the specific individual.
Medicine – remains may represent a pathological condition.
Natural history – remains may represent all humans ever.

Humans in natural history museums are essentially just another species of animal. It would be incomplete to create a display of primate skeletons without including a human (start a tally of how many museums have displays like this to see how widely accepted that view is). That's either because humans are nothing special – just another ape – or because we, as human visitors to human museums, want to see how we fit into our own zoological family. Or probably both. Such displays clearly communicate that orang-utans, gorillas, chimps and humans are essentially the same on the inside.

Humans often appear in natural history museums as part of comparative anatomy displays – designed to allow the same structures in different animals to be seen together and compared. The collection of brains on the previous page communicates how similar and different we are to pigeons, dogs, porpoises, monkeys and orang-utans, to name a few.

The human brain actually originated from University College London's Pathology Museum and the specimen's purpose has changed as a result of its move to a natural history museum – it is no longer being used as an example of a healthy brain to compare with diseased brains alongside it in a medical collection. Instead, it has become a representative of its entire species – just another animal.

But of course in the eyes of the law we are not just another animal. The ethics and morals around the treatment of human remains are likely linked to religious beliefs and traditions relating to the dead. It was interesting, then, when I oversaw a project that sought museum visitors' views on this topic. We asked the question 'Should human and animal remains be treated differently in natural history museums?' There were nearly six times as many responses saying that displaying human remains is acceptable as those who felt humans should be treated differently.

Humans *are* animals, and as long as the specimens have been acquired ethically (as required by the Human Tissues Act), I think human remains should be included in zoology displays wherever they can be instructive.

One aim of museums is to help people understand their place in the world. I think that encountering our own species in a natural history museum helps us to see how we fit into the animal kingdom. As a result, visitors might feel special because our brains are big, or be humbled by how similar we mammals are on the inside. Or they might try to interpret it in one of the modes of another discipline, and ask whose brain it is.

Treating humans as 'just another animal' is commonplace in natural history museums. Pictured here is a chimp, an orang-utan, a human and a gorilla.

100 Antarctic Limpets

BRITISH ANTARCTIC SURVEY SAMPLE

The Importance Of Natural History Collections

There is much more to a natural history museum than meets the eye, and that's mostly because relatively tiny proportions of their collections

are on display – typically less than one percent. The vast majority of specimens were not intended for public enjoyment, and that includes the last of my 100 objects. With it, I'd like to look beyond the display case and state explicitly – in case it hasn't come across already – that natural history museums may be flawed in the ways I've explored in recent chapters, but they are critical to our understanding of the natural world.

When most people think of natural history museums they think of the public galleries, for obvious reasons. And the value of these displays are both easy to imagine and statistically proven. In 2013 I oversaw a study funded by Arts Council England for the Natural Sciences Collections Association which found that natural history was the most popular gallery type in museums with mixed collections. Ahead of art, social history, archaeology and everything else, natural history galleries were the most loved and the most visited (but seldom the most financially supported).

Aside from the enjoyment they bring, museums generally are a force for good in society. They contribute to social cohesion and decrease barriers to specialist knowledge. They are also significant agents in stimulating creativity, which could hardly be more important. Colleagues in my department at University College London are investigating the benefits to people's health and wellbeing from engaging with museums. But what I want to talk about here is the world-changing science that happens in natural history museums behind the scenes. This is something that I think many natural history museums could be more successful at using their galleries to communicate.

As well as being visitor attractions, many museums are also research institutes – there may be tens or hundreds of scientists working behind the scenes, using the stored collections to make discoveries about biodiversity, evolution and how ecosystems work, among many other things. If they don't have their own researchers on staff, most museums with natural history collections hold the bulk of their collections off display for researchers to visit.

There are many kinds of research that can be done with museum specimens, not least because museums hold type specimens: the physical definitions of each species (see chapter 88). Specimens are not just an example of a species, they are an example of a species at a particular place at a particular time. As crucial as the object itself are the data that go with it: where it was collected, when, and by whom. These pieces of information make museum specimens incredibly valuable. The limpets on the previous

The Institute of Oceanographic Sciences (now the National Oceanography Centre) have been amassing 'The Discovery Collection' since 1925. The original purpose of this survey was for whaling research, so they trawled for huge numbers of samples of krill – the principal food of whales. As such, they can tell us about the oceans before the whaling industry brought about the collapse of global whale populations.

page, for example, were collected at King Edward Point in South Georgia in the Southern Atlantic Ocean on 13th November 1969, by the British Antarctic Survey.

One of the most critical kinds of research that natural history collections can be used for is investigating how ecosystems and their ecological communities change over time. That specimen is physical evidence that this species of limpet was at that site on that date. Furthermore, analysis of the make-up of the specimens could tell us about the chemistry of the ocean they lived in. Understanding the impact of climate change, pollution, ocean acidification and ecosystem collapse are essential to our ability to mitigate these global challenges, and specimens like this can provide invaluable data.

Every specimen in a museum which has provenance data can help answer questions like this. The past couple of centuries over which today's museums have come into being coincide with the extraordinary impacts humans have wrought on the world. Information about what lived where when, as well as chemical signals that can be retrieved from historic collections, are a huge line of evidence in understanding global problems.

Perhaps the most famous example is the unintended role of the pesticide DDT on bird populations. By comparing the thickness of eggshells in museums collected in the nineteenth and twentieth centuries, researchers could demonstrate that DDT was causing birds to lay thin eggs, so they smashed when the birds sat on them to incubate. This led to a ban on the insecticide and a recovery of bird populations.

The limpets pictured on p. 315 are one data point on a long running survey. Different scientific organisations have been sampling at King Edward Point since 1925. As such the specimens from different points in the long term sampling series can be compared over time. It isn't enough just to take one sample at one time.

Much of what we now know about the way that the natural world works has been learned from specimens in museums, indeed many of the stories in this book rely on discoveries from such collections. Arguably more than any other museum discipline, natural history collections make essential contributions to our understanding of our planet, answering questions that will genuinely impact our chance of surviving the changes our species is bringing about.

Natural history museums have their quirks, but what they hold is countless stories, and immeasurable value. They are so much more than just the displays.

Selected References and Further Reading

Understanding Diversity: Putting Animals In Order

Dunn, C.W., Giribet, G., Edgecombe, G.D. and Hejnol, A. (2014), 'Animal phylogeny and its evolutionary implications', *Annual Review of Ecology, Evolution, and Systematics*, vol. 45, pp. 1–651
Pechenik, J.A., *Biology of the Invertebrates, 6th Edition* (McGraw-Hill: New York, 2010)
Piper, R., *Animal Earth: The Amazing Diversity of Living Creatures* (Thames and Hudson: London, 2015)
Ruppert, E.E., Fox, R.S. and Barnes, R.D., *Invertebrate Zoology: A Functional Evolutionary Approach, 7th Edition* (Thomson-Brooks/Cole: Belmont, 2004)
The University of California Museum of Palaeontology's online exhibits has a fantastic overview of many taxonomic groups: www.ucmp.berkeley.edu/exhibits/

4. Comb jelly
Moroz, L.L. et al. (2014), 'The ctenophore genome and the evolutionary origins of neural systems', *Nature*, vol. 510, pp. 109–114
Singer, E., 'Did neurons evolve twice?', *Qanta Magazine*, Published online 25 March 2015
Zimmer, C., 'Strange findings on comb jellies uproot animal family tree', *National Geographic*, Published online 21 May 2014
5. Parasitic flatworm
O'Donoghue, P., Para-site website (The University of Queensland: Brisbane, 2010)
6. Bootlace worm
Gittenberger, A. and Schipper, C. (2008), 'Long live Linnaeus, *Lineus longissimus* (Gunnerus, 1770) (Vermes: Nemertea: Anopla: Heteronemertea: Lineidae), the longest animal worldwide and its relatives occurring in the Netherlands', *Zoologische Mededelingen*, vol. 82, pp. 59–63
7. King ragworm
Purschkea, G. et al. (2006), 'Photoreceptor cells and eyes in Annelida', *Arthropod Structure & Development*, vol. 35, pp. 211–230
9. Bryozoan
Eldredge, N., *Eternal Ephemera: Adaptation and the Origin of Species from the Nineteenth Century Through Punctuated Equilibria and Beyond* (Columbia University Press: New York, 2015)

Life's Turning Points: How Did We Get Here?

Benton, M.J., *Vertebrate Palaeontology, 3rd Edition* (Blackwell Science Ltd, 2005)
Janvier, P., *Early Vertebrates* (Clarendon Press: Oxford, 1996)
Kardong, K.V., *Vertebrates: Comparative Anatomy, Function, Evolution, 5th Edition.* (McGraw-Hill: New York, 2009)

20. *Ottoia* the penis worm
Dodd, M.S. et al. (2017), 'Evidence for early life in Earth's oldest hydrothermal vent precipitates', *Nature*, vol. 543, pp. 60–64

21. Jawless fishes
    Ota, K.G. and Kuratani, S. (2007), 'Cyclostome embryology and early evolutionary history of vertebrates', *Integrative & Comparative Biology*, vol. 47, pp. 329–337
22. Cookie-cutter shark
    Milius, S. (1998), 'Glow-in-the-dark shark has killer smudge', *Science News*, vol. 154, p. 70
24. Coelacanth
    Ahlberg, P.E. (1992), 'Coelacanth fins and evolution', *Nature*, vol. 358, p. 459

Natural Histories: How Does Evolution Work?

IUCN 2016. *The IUCN Red List of Threatened Species.* Version 2016–3

Krebs, J.R. and Davies N.B., *An Introduction to Behavioural Ecology* (Blackwell Scientific Publications: 1987)

Pough, F.H, Janis, C.M. and Heiser, J.B. *Vertebrate Life, 9th Edition* (Prentice Hall: Upper Saddle River, 2002)

Van Dyck, S., Gynther, I. and Baker, A., *Field Companion to the Mammals of Australia* (New Holland Publishers: Sydney, 2013)

Van Dyck, S. and Strahan, R., *The Mammals of Australia, 3rd Edition* (New Holland Publishers: Sydney, 2008)

31. Giant deer
    Gill, A. and West, A., *Extinct* (Channel 4 Books: London, 2001)
32. Lesser bird-of-paradise
    Bowman, J. (Executive Producer) Birds-of-Paradise Project website, Cornell University www.birdsofparadiseproject.org/
33. Garden snail
    Koene, J.M. (2006), 'Tales of two snails: Sexual selection and sexual conflict in *Lymnaea stagnalis* and *Helix aspersa*', *Integrative & Comparative Biology*, vol. 46, pp. 419–429
34. Bed bugs
    Stutt, A.D. and Siva-Jothy, M.T. (2001), 'Traumatic insemination and sexual conflict in the bed bug *Cimex lectularius*', *PNAS*, vol. 98, pp. 5683–5687
35. Seahorses
    Jones, A.G. and Avise, J.C. (2001), 'Mating systems and sexual selection in male-pregnant pipefishes and seahorses: Insights from microsatellite-based studies of maternity', *Journal of Heredity*, vol. 92, pp. 150–158
36. Gorillas
    Harcourt, A.H. et al. (1981), 'Testis weight, body weight and breeding system in primates', *Nature*, vol. 293, pp. 55–57
37. Walrus
    Fay, F.H. (1982), 'Ecology and biology of the Pacific walrus, *Odobenus rosmarus divergens* Illiger'. *North American Fauna*, vol. 74, pp. 1–279
38. Dugong
    Domning, D.P. (2001), 'The earliest known fully quadrupedal sirenian', *Nature*, vol. 413, pp. 625–627
39. Box jellyfish
    Garm, A., Oskarsson, M. and Nilsson, D. (2011), 'Box jellyfish use terrestrial visual cues for navigation', *Current Biology*, vol. 21, pp. 798–803
    Yong, E. 'Why box jellyfish always have four eyes on the sky', *Discover*, Published online 28 April 2011
40. Striped possum
    Weisbecker, V. and Goswami, A. (2010), 'Brain size, life history, and metabolism at the marsupial/placental dichotomy', *PNAS*, vol. 107, pp. 16216–16221

44. Amphisbaenian – a 'worm-lizard'

Longrich N.R., et al. (2015), 'Biogeography of worm lizards (Amphisbaenia) driven by end-Cretaceous mass extinction', *Proc. R. Soc. B*, vol. 282, pp. 2014–3034

47. Crab-eating macaque

Liedigk, R., et al. (2015), 'Mitogenomic phylogeny of the common long-tailed macaque (*Macaca fascicularis fascicularis*)', *BMC Genomics*, vol. 16, p. 222

Smith, D.G. et al. (2014). 'A genetic comparison of two alleged subspecies of Philippine cynomolgus macaques', *American Journal of Physical Anthropology*, vol. 155, pp. 136–148

48. *Megaladapis*: The koala lemur

Martin, R.D. (2000), 'Origins, diversity and relationships of lemurs', *International Journal of Primatology*, vol. 21, pp. 1021–1049

49. Hoolock gibbon

Michilsens, F. et al. (2009), 'Functional anatomy of the gibbon forelimb: Adaptations to a brachiating lifestyle', *Journal of Anatomy*, vol. 215, pp. 335–354

51. Hippopotamus

Eltringham, S.K., *The Hippos* (Academic Press: London, 1999)

53. Gaboon viper

Bramble, D.M. and Wake, D.B. 'Feeding mechanisms of lower tetrapods', in Hildebrand, M. et al. (eds), *Functional Vertebrate Morphology* (Harvard University Press: Cambridge, 1985)

54. Narwhal

Nweeia, M.T. et al. (2014), 'Sensory ability in the narwhal tooth organ system', *The Anatomical Record*, vol. 297, pp. 599–617

55. Bushmaster

Gracheva, E.O. et al. (2010), 'Molecular basis of infrared detection by snakes', *Nature*, vol. 464, pp. 1006–1011

56. Electric ray

Kramer, B. (1996), 'Electroreception and communication in fishes', *Progress in Zoology*, vol. 42

57. Cuttlefish

Crook, A.C., Baddeley, R. and Osorio, D. (2002), 'Identifying the structure in cuttlefish visual signals', *Phil. Trans. R. Soc. Lond. B*, vol. 357, pp. 1617–1624

Hanlon, R.T. and Messenger, J.B. (1988), 'Adaptive coloration in young cuttlefish (*Sepia officinalis* L.): The morphology and development of body patterns and their relation to behaviour', *Phil. Trans. R. Soc. Lond. B*, vol. 320, pp. 437–487

59. Caterpillar with parasitoid fungus

Roy, H.E. et al. (2006), 'Bizarre interactions and endgames: Entomopathogenic fungi and their arthropod hosts', *Annual Review of Entomology*, vol. 51, pp. 331–357

Taylor, P.D. and O'Dea, A., *A History of Life in 100 Fossils* (Natural History Museum: London, 2014)

Yanoviak, S.P. (2008), 'Parasite-induced fruit mimicry in a tropical canopy ant', *The American Naturalist*, vol. 171, pp. 536–544

60. Bumblebee

Clarke, D. et al. (2013), 'Detection and learning of floral electric fields by bumblebees', *Science*, vol. 340, pp. 66–69

63. Horseshoe crab

Briggs, D.E.G. et al. (2012), 'Silurian horseshoe crab illuminates the evolution of arthropod limbs', *PNAS*, vol. 109, pp. 15702–15705

Switek, B., 'In evolution's race, horseshoe crabs took a slower pace', *Wired*, Published online 22 November 2011

64. New Zealand native frog

EDGE species account, 'Archey's frog' (accessed 14 February 2017) www.edgeofexistence.org/amphibians/species_info.php?id=546

Waldman, B. (2011), 'Brief encounters with Archey's frog', *FrogLog*, vol.99, pp. 39–41

65. Kangaroo and 66. Koala

Bennett, C.V. and Goswami A. (2011), 'Does reproductive strategy drive limb integration in marsupials and monotremes?', *Mammal Biology*, vol. 76, pp. 79–83

Bennett, C.V. and Goswami A. (2013), 'Statistical support for the hypothesis of developmental constraint in marsupial skull evolution', *BMC Biology*, vol. 11, p. 52

Jackson, S. and Vernes, K., *Kangaroo: Portrait of an Extraordinary Marsupial* (Allen & Unwin: Crows Nest, 2010)

McHugh, E., *1606: An Epic Adventure* (UNSW Press: Sydney, 2006)

Sears, K.E. (2004), 'Constraints on the morphological evolution of marsupial shoulder girdles', *Evolution*, vol. 58, pp. 2353–2370

Seymour, J. (1980), 'The small marsupials', *ECOS*, vol. 26, pp. 3–9

68. Portuguese man o' war

Dunn, C., siphonophores.org (accessed 14 February 2017)

69. Mantis shrimp

Telis, G., 'Mantis shrimp smash!' *Science*. Published online 7 June 2012

Thoen, H.H. et al. (2014), 'A different form of color vision in mantis shrimp', *Science*, vol. 343, pp. 411–413

deVries, M.S., Murphy, E.A.K. and Patek, S.N. (2012), 'Strike mechanics of an ambush predator: The spearing mantis shrimp', *Journal of Experimental Biology*, vol. 215, pp. 4374–4384

71. Olm

Laudet, V. (2011), 'The origins and evolution of vertebrate metamorphosis', *Current Biology*, vol. 21, pp. R726–737

73. Aphids

Hales, D.F. et al. (2002), 'Lack of detectable genetic recombination on the X chromosome during the parthenogenetic production of female and male aphids', *Genetical Research*, vol. 79, pp. 203–209

Judson, O., *Dr. Tatiana's Sex Advice to All Creation* (Metropolitan Books: New York, 2002)

van der Kooi, C.J and Schwande, T. (2015), 'Parthenogenesis: Birth of a new lineage or reproductive accident?', *Current Biology*, vol. 25, pp. R659–661

77. Elephant bird

Lomolino, M.V. (2005), 'Body size evolution in insular vertebrates: Generality of the island rule.' *Journal of Biogeography*, vol. 32, pp. 1683–1699

Mitchell, K.J. et al. (2014), 'Ancient DNA reveals elephant birds and kiwi are sister taxa and clarifies ratite bird evolution', *Science*, vol. 344, pp. 898–900

79. Tasmanian devil

Hawkins, C.E. et al. (2008), *Sarcophilus harrisii*. The IUCN Red List of Threatened Species 2008: e.T40540A10331066. Downloaded on 16 February 2017

Pye, R. et al. (2016), 'Demonstration of immune responses against devil facial tumour disease in wild Tasmanian devils', *Biology Letters*, vol. 12, 20160553

Epstein, B. (2016), 'Rapid evolutionary response to a transmissible cancer in Tasmanian devils', *Nature Communications*, vol. 7, 12684

80. Tree frogs

Amphibian Ark (2016), *Chytrid Fungus* www.amphibianark.org/the-crisis/chytrid-fungus/ (accessed 16 February 2017)

81. Marine iguana

Cai, W. et al. (2014), 'Increasing frequency of extreme El Niño events due to greenhouse warming', *Nature Climate Change*, vol. 4, pp. 111–116

Gunderson, A.R. and Stillman, J.H. (2015), 'Plasticity in thermal tolerance has limited potential to buffer ectotherms from global warming', *Proc. R. Soc. B*, vol. 282, 20150401

82. Crescent nailtail wallaby

Australian Wildlife Conservancy (2013), 'Stopping the slaughter: Fighting back against feral cats', *Wildlife Matters*, Summer 2012–13

Doherty, T.S. et al. (2016), 'Invasive predators and global biodiversity loss', *PNAS*, vol. 113, pp. 11261–11265

Johnson, C.N. and Isaac, J.L. (2009), 'Body mass and extinction risk in Australian marsupials: The "Critical Weight Range" revisited', *Austral Ecology*, vol. 34, pp. 35–40

Legge, S. et al. (2016), 'Enumerating a continental-scale threat: How many feral cats are in Australia?', *Biological Conservation*, vol. 206, pp. 293–303

83. Trilobites

Labandeira C.C. and Sepkoski J.J. (1993), 'Insect diversity in the fossil record', *Science*, vol. 261 pp. 310–315

85. Collie and King Charles spaniel

Clutton-Brock, J., *A Natural History of Domesticated Mammals, 2nd Edition* (Cambridge University Press: Cambridge, 1999)

Driscoll, C.A., Macdonald, D.W. and O'Brien, S.J. (2009), 'From wild animals to domestic pets, an evolutionary view of domestication', *PNAS*, vol. 106, pp. 9971–9978

Thalmann O. et al. (2013), 'Complete mitochondrial genomes of ancient canids suggest a European origin of domestic dogs', *Science*, vol. 342, pp. 871–874

Wang, X. and Tedford, R.H., *Dogs: Their Fossil Relatives and Evolutionary History* (Columbia University Press: New York, 2008)

86. Domestic pig

Mauriès, P., *Cabinets of Curiosities* (Thames and Hudson: London, 2013)

88. James Bond's hutia

Marshall S.A. and Evenhuis N.L. (2015), 'New species without dead bodies: a case for photo-based descriptions, illustrated by a striking new species of *Marleyimyia* Hesse (Diptera, Bombyliidae) from South Africa', *ZooKeys*, vol. 525, pp. 117–127

Naish, D., 'Animal species named from photos', *Scientific American*, Published online 3 February 2017

Turvey S. et al. (2015), 'A new subspecies of hutia (*Plagiodontia*, Capromyidae, Rodentia) from southern Hispaniola', *Zootaxa*, vol. 3957, pp. 201–214

91. *Hypsilophodon*

Taylor, M., 'How long was the neck of *Diplodocus*?', *Sauropod Vertebra Picture of the Week*, Published online 19 May 2011

92. Dodos

Fuller, E., *Dodo. A Brief History* (Collins: London, 2002)

Parish, J.C. (2015), 'A catalogue of specimens of the dodo (*Raphus cucullatus*) and the Solitaire (*Pezophaps solitaria*) in collections worldwide', *The Dodologist's Miscellany*, Published online 26 January 2015

96. Domestic cat

Machin, M. (2008), 'Gender representation in the natural history galleries at the Manchester Museum', *Museum and Society*, vol. 6, pp. 54–67

100. Antarctic limpets

Jenkins, S., Lisk, J. and Broadley A. (2013), 'The popularity of museum galleries'

Acknowledgements

This book has been an absolute joy to write, and I'd really like to thank the many people who've been such a help along the way. I am extremely grateful to my editors at The History Press: Sophie Bradshaw for inviting me to write it; and Christine McMorris for producing the final publication.

Given the range of animal groups and topics covered by this book, I have been very lucky that true specialists were able to look over the chapters relating to their area of the animal kingdom and provide invaluable comments. Miranda Lowe of the Natural History Museum was extraordinarily generous to look at all of the invertebrates. Mark Carnall from the Oxford University Museum of Natural History took on proofing the Displaying Nature Section, as well as suggesting a couple of the objects, and I'd also like to thank him for all the conversations through the 10 years we worked together that honed a lot of these thoughts. Dr Marc Jones and Dr Thomas Halliday were extremely helpful in looking over sizeable chunks of the reptile and amphibian, and mammal chapters respectively.

I am very grateful to the following, who were kind enough to proof-read chapters linked to their own research: Carla Bardua, Prof Helen Chatterjee (who also convinced me to explore the wonders of gibbons further), Dr Ryan Felice, Prof Anjali Goswami, Dr Rodrigo Hamede, Prof Kate Jones, Dr Louise Martin, Dr Hugh McGregor, Toby Nowlan, Marcela Randau, Dr Katherine Tuft and Dr Akinobu (Aki) Watanabe.

Finally, my mum Sue – who read the whole thing (in parts more than once) – and my dad Dave were of great help in checking the sense and readability of what I wrote – thanks!

I'd like to thank my team at the Grant Museum of Zoology, UCL, who have been supportive throughout (as well as caring for the collections on which this book is based): conversations with Tannis Davidson, Emilia Kingham, Will Richard, Rowan Tinker, Dean Veall and Paolo Viscardi have helped me shape my thoughts around the objects. Rowan was particularly helpful in sifting through the images for the book as well as providing some background research on our island mice collection.

Oliver Siddons provided much of the object photography, as has Tony Slade. I'd also like to thank Jamie Coleman for initial help with the contractual logistics of the project.

A number of other people suggested lines of investigation, topics to write about and answered queries. They were: Paul Barrett, Joe Cain, Erica McAlister, @Chris_Manias, @doggomoose, @gsciencelady, @joeyrclarke, @_mammaliana_, @matthewcobb (particularly for pointing me to @DrRossPiper's book), and @zoologyrohan.

Image Credits

© Diego Delso, delso.photo, Licence CC-BY-SA
p. 257

© Jack Ashby
p. 252

© Olgavisavi/Dreamstime
p. 36

© Oxford University, Oxford University Museum of Natural History
p. 293

© Paul Nicklen
p. 128

Shutterstock.com
Andrea Izzotti © p. 188; Anna Veselova © p. 138; Arrixx © p. 42; Catmando © p. 87; evantravels © p. 24; Francesco de Marco © p. 203; Graeme Shannon © p. 171; Hiso © p. 122; John Carnemolla © p. 218; Kiki Dohmeier © p. 126; Linda Bucklin © p. 180; Matt Jeppson © p. 141; Procy © p. 168; Setaphong Tantanawat © p. 165; Szefei © p. 115; Vladimir Wrangel © p. 184

© UCL Grant Museum of Zoology
pp. 20, 21, 41, 55, 88, 90, 113, 136, 166, 174, 244, 287

© UCL Grant Museum of Zoology/Fred Langford Edwards
pp. 10, 17, 68, 121, 154, 169, 178–9, 204, 225, 228, 239, 247, 258, 265, 300, 307

© UCL Grant Museum of Zoology/Matt Clayton
pp. 270, 294, 296

© UCL Grant Museum of Zoology/Oliver Siddons
pp. 18, 23, 30, 34, 38, 45, 50, 52, 56, 58, 62, 64, 73, 76–7, 80, 82, 86, 91, 94, 97, 101, 107, 110, 116, 119, 124, 127, 130, 134, 139, 142, 146, 148, 151, 160, 163, 172, 176, 181, 186, 190, 192, 196, 198, 200, 206, 208, 219, 222, 227, 230, 232, 233, 236, 242, 250, 253, 255, 261, 263, 267, 273, 276, 279, 282, 285, 297, 299, 312, 314, 315, 316

© UCL Grant Museum of Zoology/Tony Slade
pp. 32, 35, 157, 183, 201, 210, 213, 215, 216, 241, 288, 291, 302, 303, 310

© UCL Grant Museum of Zoology, UCL Centre for Digital Humanities/Kira Zumkley
pp. 48

Wikimedia Commons
Acrocynus © p. 246; B.H. Michael © p. 162; David P. Hughes, Maj-Britt Pontoppidan © p. 194; Dmitry Bogdanov © pp. 93, 96; Karen N. Pelletreau et al. © p. 191; Luis Fernández García © p. 60; Mcy Jerry © p. 102; MedievalRich © p. 235; Smokeybjb © p. 84

Index

Note: *italicised* page numbers indicate illustrations